Mario Pagliaro

Nano-Age

Related Titles

Garcia-Martinez, J. (ed.)

Nanotechnology for the Energy Challenge

2010
ISBN: 978-3-527-32401-9

Geckele, K. E., Nishide, H. (ed.)

Advanced Nanomaterials

2010
ISBN: 978-3-527-31794-3

Amabilino, D. B. (ed.)

Chirality at the Nanoscale

Nanoparticles, Surfaces, Materials and more

2009
ISBN: 978-3-527-32013-4

Rubahn, H.-G.

Basics of Nanotechnology

2008
ISBN: 978-3-527-40800-9

Borisenko, V. E., Ossicini, S.

What is What in the Nanoworld

A Handbook on Nanoscience and Nanotechnology

2008
ISBN: 978-3-527-40783-5

Balzani, V., Credi, A., Venturi, M.

Molecular Devices and Machines

Concepts and Perspectives for the Nanoworld

2008
ISBN: 978-3-527-31800-1

Köhler, M., Fritzsche, W.

Nanotechnology

An Introduction to Nanostructuring Techniques

2007
ISBN: 978-3-527-31871-1

Mario Pagliaro

Nano-Age

How Nanotechnology Changes our Future

WILEY-VCH Verlag GmbH & Co. KGaA

The Author

Prof. Mario Pagliaro
CNR
Ist. Materiali Nanostrutturati
via Ugo La Malfa 153
90146 Palermo
Italien

Cover

Provided with kind permission
from FRAUSCHER BOOTS
WERFT GmbH & Co KG

All books published by Wiley-VCH are carefully produced. Nevertheless, authors, editors, and publisher do not warrant the information contained in these books, including this book, to be free of errors. Readers are advised to keep in mind that statements, data, illustrations, procedural details or other items may inadvertently be inaccurate.

Library of Congress Card No.: applied for

British Library Cataloguing-in-Publication Data
A catalogue record for this book is available from the British Library.

Bibliographic information published by the Deutsche Nationalbibliothek
The Deutsche Nationalbibliothek lists this publication in the Deutsche Nationalbibliografie; detailed bibliographic data are available on the Internet at <http://dnb.d-nb.de>.

© 2010 WILEY-VCH Verlag GmbH & Co. KGaA, Boschstr. 12, 69469 Weinheim, Germany

All rights reserved (including those of translation into other languages). No part of this book may be reproduced in any form – by photoprinting, microfilm, or any other means – nor transmitted or translated into a machine language without written permission from the publishers. Registered names, trademarks, etc. used in this book, even when not specifically marked as such, are not to be considered unprotected by law.

Cover Design   Adam Design, Weinheim
Typesetting   Toppan Best-set Premedia Limited, Hong Kong
Printing and Binding   Strauss GmbH, Mörlenbach

Printed in the Federal Republic of Germany
Printed on acid-free paper

ISBN: 978-3-527-32676-1

"This book is dedicated with love to my sister Annalisa"

Contents

Preface *IX*
About the Author *XIII*

1 **Capturing Sun's Energy** *1*
1.1 Solar Power: Now *1*
1.2 Never Trust the Skeptics *2*
1.3 Solar Power for the Masses *6*
1.4 Why Nanoscience is Relevant to the Solar Energy Industry *9*
1.5 Expanding the Solar Business *15*
1.6 Solar Hydrogen from Water *19*
References *24*

2 **From Chemistry to Nanochemistry** *27*
2.1 Why Small is Different *27*
2.2 Nanochemistry, the Chemical Approach to Nanotechnology *29*
2.3 An Insight into Chemical Methodology *31*
2.4 Making Nanomaterials *39*
References *43*

3 **Storing and Supplying Clean Energy** *45*
3.1 Ending the Era of the Internal Combustion Engine *45*
3.2 Nanotechnology-Based Batteries *49*
3.3 Biological Fuel Cells *54*
3.4 Fuel Cells for the People *58*
References *63*

4 **Catalysis: Greening the Pharma Industry** *65*
4.1 Pharma: An Industry to Be Cleaned Up *65*
4.2 Sol-Gel Catalysts: Philosopher's Stones *69*
4.3 Biogels: Marriage of Glass and Life *76*
4.4 Nanocatalysts: Abating Polluting Emissions and Product Contamination *81*
References *85*

Nano-Age: How Nanotechnology Changes our Future. Mario Pagliaro
© 2010 WILEY-VCH Verlag GmbH & Co. KGaA, Weinheim
ISBN: 978-3-527-32676-1

5	**Organically Doped Metals** *87*
5.1	A Watershed Development in Science *87*
5.2	The New Reactivity of Metal-Entrapped Molecules *90*
5.3	Two-for-One-Catalyst *93*
5.4	Chiral Metals *95*
	References *98*

6	**Protecting Our Goods and Conserving Energy** *101*
6.1	Multifunctional Nanocoatings *101*
6.2	Multifunctional Textiles *109*
6.3	Protecting Cultural Heritage *110*
6.4	Protecting Goods from Light *111*
	References *117*

7	**Better Medicine Through Nanochemistry** *119*
7.1	Nanomedicine *119*
7.2	Hemostasis: Change in Surgery and Emergency Medicine *122*
7.3	Biogels: Biotechnology Made Possible *123*
7.4	Small is Beautiful? Nanotech Cosmetics *125*
7.5	Nanotechnology in Orthopedics *129*
7.6	A Hybrid, Welcome Science *131*
	References *136*

8	**Getting There Cleanly** *139*
8.1	Why Sustainable Nanotechnology? *139*
8.2	Regulating Nanomaterials *141*
8.3	Greening Nanomaterials *143*
8.3.1	Cleaning Up Water *144*
8.3.2	Biocompatible Coatings *145*
8.3.3	Green Metal Nanoparticles *145*
8.4	Understand the Risks and Minimize Them *146*
8.5	Communicating the Nanotech Risk *148*
8.5.1	Cultural Message Framing *150*
8.5.2	Contextualization *150*
	References *151*

9	**Managing (Nano)innovation** *153*
9.1	Scholars, and not Researchers *153*
9.2	Renewing Management and Scientific Education *155*
9.3	Nexus of the Sciences *159*
9.4	In Praise of Scientific Culture *162*
9.5	Communicating Nanochemistry *164*
	References *170*

Index *171*

Preface

While sitting after lunch with the economist Loretta Napoleoni in front of Sicily's splendid sea, Loretta made the rather straightforward statement that *"Nanotechnology will save us from a global collapse."* She went on explaining her argument for another two hours.

When a field of science becomes the subject of discussion for non-scientists, and for members of the economic elite in particular, I become interested. And so probably are readers in the political, economic, and media communities out there.

When most people think of nanotechnology–if they think of nano technology at all–common images are futuristic tiny robots, performing advanced surgery or being deployed on the battle ground. Well, forget about this hype.

Is there anything really relevant for society in the hype surrounding "nanotechnology" that is creating and manipulating objects whose functions are due to dimensions or components at a billionth of a meter scale?

Investors and wise readers who remember the dot-com bubble and subsequent burst, please be patient. Predicting the future of fast-moving technical fields is difficult, but, yes, there is much of relevance for you and for society here, after the wheat is separated from the chaff.

Nanotechnology is a new and vitally important area of interdisciplinary scientific research that has gone global. It actually is materials chemistry at the nanoscale, or *nanochemistry*. Like chemistry, it has to do with how we make useful things.

Chemists have invented all manner of stuff–drugs, plastics, fertilizers, glasses, dyes, explosives, contact lenses, textiles, solar panels, ink, coatings, fuels ...–that have had such a broad impact on the everyday's life of all of us.

Now and in the near future chemists working in co-operation with other scientists will provide society with a wide variety of new (nano)materials for a myriad applications of immense practical importance, spanning the fields of chemistry and physics, materials science and engineering, biology, and medicine.

Nano-Age: How Nanotechnology Changes our Future. Mario Pagliaro
© 2010 WILEY-VCH Verlag GmbH & Co. KGaA, Weinheim
ISBN: 978-3-527-32676-1

Indeed, the boundaries that have separated these traditional chemistry disciplines in the twentieth century, and chemistry from other disciplines such as physics, biology, and engineering, have broken down to create one large multidisciplinary community with a keen scientific and technological interest in "all" aspects of the chemistry of materials at the nanoscale.

A true chemistry of materials has emerged in the last 15 years as scientists from all disciplines have learned how to synthesize and exploit new types of materials from individual or groups of nanoscale building-blocks that have been intentionally designed to exhibit useful properties with purposeful function and utility [1].

Such techniques generally rely on formulas that control the precise, bottom-up chemical assembly of molecules into geometric structures composed of many molecules. Molecular self-assembly techniques for instance now give us the unprecedented capability of designing and creating nanostructured materials with novel properties.

In practice, synthetic chemistry is now used to make *nanoscale* building blocks with controlled size and shape, composition, surface structure, and functionality that can be useful in their own right or in a self-assembled structure. In addition, since the properties of a (nano)material emerge from the composition, size, shape, and surface properties of these individual building-blocks, chemists are becoming increasingly able to synthesize, from the bottom-up, tailor-made materials.

Beyond research, powerful trends are already evident in business. Not only do a growing number of start-up companies now commercialize products obtained via the nanochemistry approach but also national laboratories, military establishments, and the very big chemical companies have entered the field and joined the race for new and exciting nanomaterials.

Trends in public and private funding of nanotechnology have evolved accordingly. Global research funding in the USA, Russia, Canada, Japan, China, India, Korea, and the EU is in the range of billions of dollars.

In *Nano-Hype* David Berube concluded that:

> "much of what is sold as 'nanotechnology' is in fact a recasting of previous materials science, which is leading to a nanotech industry built solely on selling nanotubes, nanowires, and the like which will end up with a few suppliers selling low margin products in huge volumes" [2].

Or, even better said:

> "In the 1990s keywords for research projects were 'green', 'environmentally friendly' and 'high throughput'. Then we had 'supramolecular', 'miniaturisation' and 'interface'. And now, it is 'nano' everywhere. But what is nanotechnology? We need to stop abusing this word 'nano'. If we define 'nano' by scale alone, then most chem-

ists are working on 'nanochemistry'. Indeed I found that some scientists have started to label anything small as 'nano'. Three years ago, I went to a nanobiotechnology meeting at London and found about half of the presentations have very little to do with real nanotechnology. One of the keynote lectures was on measuring movements of wings of insects at the nanometre scale. Others were just pictures of nanocrystals of some biological samples (we called them 'crystallites' in the past) … For example, there was a presentation on enzymatic catalysis and enzyme inhibitors – while an enzyme molecule is around a few nanometres wide, are they really 'nanocatalysts'? Or is this actually the molecular biochemistry that we have been learning for decades?" [3]

Yet, although there has been much sensationalism about the potential applications of nanotechnology, nanomaterials are up to having a revolutionary impact in several fields of enormous societal relevance.

One example of such disruptive technology is the generation of clean electricity from solar energy with new photovoltaic solar cells that can produce electricity at $0.10 per watt, that is, the price of electricity generated by burning coal, the cheapest fossil fuel still massively used worldwide.

Such an advance relies entirely on the ability of the company to architect and assemble, by fast roll-to-roll printing, nanocrystals of an inorganic material denoted with the acronym CIGS, on a scale in the 1–100 nm range, namely, at the length scale at which the relevant photovoltaic quantum-physics occurs.

Several other advanced nanotechnologies are about to reach the market. Consequently, even if numerous researchers and entrepreneurs abuse the term "nano" to access funding, there will not be a nano bubble – especially after the global financial crisis due to "subprime" loans. When the crisis is over, investors will wisely select technologies and companies in which to invest. And nanotech companies, manufacturing brand new products with new functionalities, will be there to benefit.

Nor shall we commit the same mistakes made with the unregulated use of chemicals in the first half of the twentieth century. The serious questions about the health, environmental, and social impacts of this powerful new technology are being, and will be, dealt with before large-scale commercialization, and nanomaterials will be regulated as new chemicals.

As the economic, social, and environmental problems associated with the sustainability crisis are becoming evident on a global scale, we urgently need advanced technologies that can drastically reduce carbon dioxide emissions in the atmosphere and, at the same time, increase productivity of the processes we use to manufacture goods and produce the services capable of satisfying the (growing) needs of a global population in rapid growth.

Putting the discussion in this context of crisis and using a critical approach, this book shows how and why nanotechnology holds great promise for addressing these needs. Using plain language and plenty of examples of emerging technologies and innovations, we aim to provide readers with a unified and essential picture of nanotechnology and its impact.

Accordingly, readers of this book may include decision makers at all levels, from managers to politicians, including media professionals and educators. Nanoscale technology innovation will involve virtually every major industry. Hence, all these professionals need to increase their basic knowledge of nanoscale technology, especially of how nano-innovation will transform technologies and markets, and open new growth opportunities in all countries. Finally, the primary aim of this book is to promote action based on such knowledge: what to do and whom to contact to start benefiting from the fruits of nano-innovation.

Palermo, September 2009 *Mario Pagliaro*

References

1 Ozin, G., Aresenault, A., and Cademartiri, L. (2009) *Nanochemistry*, RSC Publishing, Cambridge.
2 Berube, D. (2006) *Nano-Hype. The Truth Behind the Nanotechnology Buzz*, Prometheus, New York.
3 Yiu, H. (2007) Defining nano. *Chemistry World*, **4**, 41–42..

About the Author

Mario Pagliaro (b. Palermo, 1969) is a research chemist and management educator based in Palermo at Italy's CNR, where he leads a research group and Sicily's Photovoltaics Research Pole. His research focuses on the development of functional materials for various uses and operates at the boundaries of chemistry, biology, and materials science.

Prof. Pagliaro is author of ten books, including the scientific bestseller *Flexible Solar Cells* (Wiley-VCH, 2008), he has co-authored international patents and a large number of scientific papers. In 2009 he chaired the 10th International FIGIPAS Meeting in Inorganic Chemistry in Palermo.

Since 2004 he has organized the prestigious Seminar "Marcello Carapezza." In 2008 he was invited to give the "John van Geuns" Lecture at the University of Amsterdam. In 2005 he was appointed *maître de conférences associé* at the Montpellier Ecole Nationale Supérieure de Chimie.

Between 1998 and 2003 he led the management educational center, Quality College del CNR established a research group that currently collaborates with researchers in ten countries (www.qualitas1998.net).

1
Capturing Sun's Energy

1.1
Solar Power: Now

Electricity is silent, clean, and easily transported and converted into work. Unsurprisingly, therefore, electric power is the most useful and desirable form of energy available to modern society. Yet, exactly like hydrogen, electricity is an energy *vector* and not an energy source. This means that we need to produce electric power by converting primary energy sources into electric power. At present, besides a 16% share from nuclear fission (www.world-nuclear.org), we produce electricity mainly by burning hydrocarbons and, sadly, much cheaper coal. For example, still, in 2006 nearly half (49%) of the 4.1 trillion kilowatt-hours (kWh) of electricity generated in the USA used coal as its energy source (www.eia.doe.gov).

Overall, presently, close to 80% of the energy supply worldwide is based on fossil fuels like coal, oil, and gas [1]. Over the next decade, China alone will need to add some 25 GW of new capacity each year to meet demand—equivalent to one large coal power station every week. Coal, unfortunately, contains mercury and, along with production of immense amounts of climate-altering CO_2, its combustion is causing pollution of the oceans and of the food chain. To abate emissions and stop climate change, the biggest challenge of our epoch is to obtain electricity directly from the sun.

The conclusions of the United Nations Intergovernmental Panel on Climate Change (IPCC) analyses concerning the potential impact of various scenarios of CO_2 emission trajectories (www.ipcc.ch) provide evidence that if the global average temperature increases beyond 2.5–3.0 °C serious, and likely irreversible, negative consequences to the environment will result, with a direct impact on agriculture, water resources, and human health [2]. In the absence of serious policies to reduce the magnitude of future climate changes, the globe is expected to warm by about 1–6 °C during the twenty-first century. The estimated carbon dioxide (CO_2) concentration in 2100 will lie in the range 540–970 ppm (parts per million), which is sufficient to cause substantial increases in ocean acidity as well as *irreversible* modifications to climate and nature [3].

Clearly, to satisfy our growing energy needs (a predicted *doubling* of energy demand by mid-century and a *tripling* by the end of the century) [21] and resolve the world's sustainability crisis, we urgently need to use a fraction of the immense solar energy amount that the Earth receives annually from the sun: 3.9×10^{24} joules, namely four orders of magnitudes larger than the world energy consumption in 2004 (4.7×10^{20} J), and enough to fulfill the yearly world demand of energy in less than an hour.

Realization of such a change will require a massive effort to discover and develop new technology solutions to capture vast amounts of dilute and intermittent, but essentially unlimited, solar energy to sustain our forecasted needs; and also to scale its conversion into high power densities and readily storable forms. The energy and societal challenge before us in developing CO_2-neutral, solar energy differs in two ways from past large-scale challenges, namely, in:

1) the large magnitude and relatively short time scale of the transition;
2) the cost-competitive aspect of the transition.

Solutions will be achieved by exploiting advances in nanoscale science and technology. Revolutionary nanotechnology-enabled photovoltaic materials and devices are being developed to satisfy the world demand in sustainable electricity at a price level of less than $0.10 per kWh. Similarly, nanotechnology-based disruptive devices for electric energy storage, such as new batteries and capacitors, are required to enable massive deployment of solar power. Most importantly, this transformation must be accomplished in an economically and environmentally feasible manner in order to achieve a positive global impact. Hence, the cost of battery technologies must plummet from the current $1000 per kWh to a level approaching $10 per kWh to become cost competitive.

1.2
Never Trust the Skeptics

Exactly in the same way that the Internet was not invented by taxing the telegraph, so cheap and abundant electricity from the sun will not be obtained by adding taxation on carbon dioxide emissions, but rather by inventing new, affordable clean technologies [4]. Nanoscale science and technology will play a crucial role in reaching this aim.

Solar power currently provides only a very small fraction of global electric power generation, about 12 GW of installed capacity globally, as of 2009, but newly installed capacity is growing at approximately 30% per year and

Figure 1.1
The Solucar 11 MW solar thermal plant outside Seville, Spain, produces enough electricity to power 6000 homes. CSP is a large-scale technology capable of satisfying massive electricity demand. Photograph courtesy of the BBC; reproduced with permission.

is accelerating. Concentrated solar power (CSP) has already been identified as a clean technology that can satisfy the world's rapidly growing energy demand.[1] Accordingly, investments are finally flowing and the first CSP plants, such as that in the Spain's city of Seville (Figure 1.1) serving 6000 families, are starting operation.

In Europe, France's President Sarkozy and Germany's Chancellor Merkel seem to have understood the situation. In establishing the new Union for the Mediterranean (Figure 1.2) they and all other government leaders of the member States have agreed to explore feasibility of large-scale generation of solar electricity in North Africa to supply Europe and Middle East countries through solar thermal technologies. Meanwhile, the often invoked two billion citizens of developing countries lacking access to an electric grid will start to self-generate electricity for their basic needs using photovoltaic modules whose price has dropped from almost $100 per watt in 1975 to $1.6 per watt at the beginning of 2009.

Direct conversion of the sun's radiation into electricity, namely photovoltaics (PVs), is being developed rapidly. After 40 years of losses and governmental subsidies, the $37 billion photovoltaics industry has turned into a profitable business, growing for the last decade at 30% per annum [5]. In this context, the installation of thin-film (TF) systems more than

---

1) One of the world's leading solar thermal CSP technologies in terms of reduced cost and safety is that of the US company Ausra: www.ausra.com.

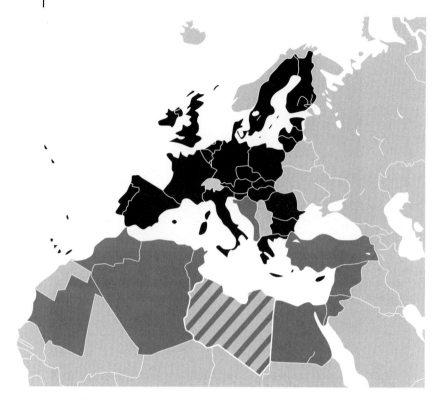

Figure 1.2
Established in 2008, the Union for the Mediterranean consists of 42 countries and has a major project: a Mediterranean Solar Plan to install concentrating solar power in the deserts and feed huge amounts of electricity to all member States, thereby ending energetic dependency on oil and natural gas. Reproduced from wikipedia.org, with permission.

doubled last year and now accounts for some 12% of solar installations worldwide [2]; the revenue market share of TFPV was expected to rise to 20% of the total PV market by 2010 (Figure 1.3) [6]. Thin-film modules are less expensive to manufacture than traditional silicon-based panels and have considerably lowered the barrier to entry into the photovoltaic energy business. From the heavy, fragile silicon panels coated with glass the sector is thus rapidly switching to thin film technologies using several different photovoltaic semiconductors.

Four years of high and increasing oil prices and the first ubiquitous signs of climate change have been enough to assist the market launch of several photovoltaic technologies based on thin films of photoactive material that had been left dormant in academic and industry laboratories for years. For example, in 2004 a leading manager in the PV industry, exploring the industry perspectives, concluded:

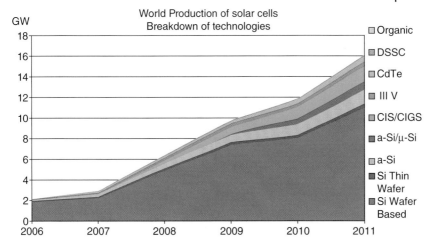

Figure 1.3
The forecast for the photovoltaic market, with a breakdown per technology, points to an annual growth rate of 70% for TFPV from 2007 to 2010 Reproduced from yole.fr, with permission.

"thin-film PV production costs are expected to reach $1 per watt in 2010, a cost that makes solar PV competitive with coal-fired electricity" [7].

However, the first thin-film solar modules profitably generating electricity for $0.99 per watt (i.e., the price of coal-fired electricity) were commercialized in late 2008 by the US company First Solar.

Now, regarding the "feed-in" tariff schemes by which, initially, governments in countries like Germany and Spain and now Italy, France, and Greece aim to incentivize production of solar electricity, the "skeptical environmentalist" Bjørn Lomborg [8] (Figure 1.4) would probably tell you that this is just another example of the poor (through an extra tax on the electric bill) financing the rich (purchasing the solar modules). But Bjørn, in this case, would be wrong.

The governments of these countries established the incentives to develop a large and modern photovoltaic industry that, thanks to the large profits fueled by the incentives, could finance innovation and lower the price of solar electricity. How right they were! Indeed, a large body of new commercial PV technologies has emerged in the last five years and last month the US-based company First Solar announced a reduction in production costs to 1\$ W^{-1} (one dollar per watt), from 5\$ W^{-1} in 2005 when the company started production of its modules based on cadmium telluride (a readily available and non-toxic inorganic salt). Cumulative

Figure 1.4 "The poor financing the rich" would say the "skeptical environmentalist" Bjørn Lomborg commenting on the "feed-in tariff" scheme adopted by Germany (and many other states). But Bjørn, in this case, would be wrong. Photograph by Emil Jupin, reproduced from lomborg.com, with permission.

production planned for 2009 by this company alone is 1 GW, that is, the power similar to that generated by a new-generation large nuclear power plant.

In management, as in politics, never trust the skeptics too much!

1.3
Solar Power for the Masses

So-called "thin film" photovoltaics are opening the route to low cost electricity. In this context, thin films are 100 nm–100 μm thick and made of organic, inorganic, and organic–inorganic solar cells deposited over rigid or flexible substrates by high-throughput (printing) technologies. If, for example, 35 μm of silicon were used to manufacture a solar cell instead of the state of the art 300 mm – such as in the case of the *SilFoil* technology based on large area, multi-crystalline silicon foils developed by NanoGram (www.nanogram.com) (Figure 1.5) – the cost of silicon-based solar modules would be below $1 W^{-1}.

While similar small start-ups eagerly compete to introduce new solar technologies, First Solar already manufactures the equivalent of one large

TECHNOLOGY OVERVIEW

A NEW PARADIGM IN SOLAR ENERGY

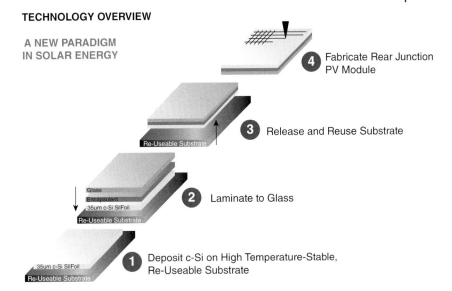

Figure 1.5 Nanogram plans to start production of its *SilFoil* modules in 2010 in a US-based production facility; 5 MW of capacity on ultra-thin crystalline silicon, which it says will reduce the cost of silicon-based solar cells to below $1 per watt by an innovative technology based on large area, multi-crystalline silicon foils. Reproduced from nanogram.com, with permission.

nuclear power station, using a deposition technique of inorganic nanocrystals of cadmium telluride. Floated on the New York Stock Exchange (NASDAQ) in November 2006 the company's stock price rapidly recovered from the financial turmoil due to the 2008 financial crisis (Figure 1.6) as production – and sales – increased 100-fold and the company has become the world's second largest in the photovoltaic business.

The above is due to technical control of the deposition process using a technique called chemical vapor deposition. Without such control, such a *disruptive* technology would not be on the market. It is disruptive because, in the face of the new threat posed by this highly competitive technology, manufacturers of traditional silicon solar panels reacted by ramping production, aiming to reduce cost by economies of scale. This, in its turn, led to oversupply of polysilicon (the raw material for Si-based solar cells) from silicon manufacturers.[2] The overall result is that, in early 2009, solar modules in Europe and in the rest of the world were selling at 1.7€ W^{-1}. Solar energy for the masses is now a reality!

2) The mean price of polysilicon to be supplied in 2009 in contracts already signed is 43% lower than the mean price of polysilicon supplied in 2008. (New Energy Finance, 2008).

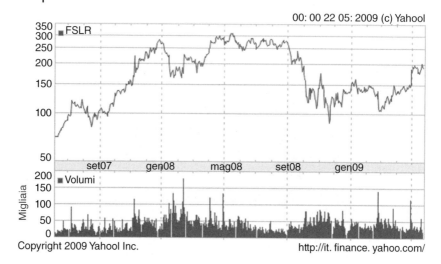

Figure 1.6
Price of First Solar stock, 2007–2009 (as of May 2009). Reproduced from google.com, with permission.

At present, most existing production tools in the solar industry have a 10–30 megawatt (MW) annual production capacity. A single machine with a gigawatt throughput would be highly desirable, leading to far higher returns on the capital invested. In late 2008, thus, Nanosolar opened a factory in California that can produce 430 MW of solar capacity a year, nearly the size of an average coal-fired power plant, by simply printing a nanoparticle ink. More precisely, the company manufactures thin film copper-indium-gallium-selenide (CIGS) solar cells by printing the active CIGS material onto mile-long rolls of thin aluminum foil (Figure 1.7), which is later cut into solar panels.

The new ink (Figure 1.8) is made of a homogeneous mix of CIGS nanoparticles stabilized by an organic dispersion. Chemical stability ensures that the atomic ratios of the four elements are retained when the ink is printed, even across large areas of deposition. This is crucial for delivering a semiconductor of high electronic quality and is in contrast to vacuum deposition processes where, due to the four-element nature of CIGS, one effectively has to "atomically" synchronize various materials sources.

Printing is a simple, fast, and robust coating process that further lowers manufacturing costs. Rolls that are meters wide and miles long are thus processed efficiently with very high throughput. Somewhat ironically, the company has been established by the heir of a wealthy German family that became rich selling, since the 1920s, electricity obtained by burning coal [9].

Figure 1.7
Nanosolar states that its 1GW CIGS coater costs $1.65 million. Working at speeds of 100 feet min^{-1} speed, the process is an astonishing two orders of magnitude more capital efficient than a high-vacuum process: a 20-times slower high-vacuum tool would have cost about ten times as much. Reproduced from nanosolar.com, with permission.

Figure 1.8
Nanosolar's nanostructure CIGS ink. The ink serves a useful purpose by effectively "locking in" a uniform distribution even across large areas of deposition. Photograph courtesy of nanosolar.com, reproduced with permission.

1.4
Why Nanoscience is Relevant to the Solar Energy Industry

From a scientific viewpoint, thin film solar cells are the result of advances in nanochemistry. Interestingly, the inventor of the silicon solar cell had already in 1954 clearly forecasted that thin films would be the configuration of forthcoming industrial cells. Indeed, it has been our chemical ability to manipulate matter on the nanoscale for *industrial* applications that has recently made possible the synthesis of the photoactive layers needed to carry out the photovoltaic conversion with the needed stability required for practical applications. Almost unnoticed among more glamorous scientific disciplines, chemistry in the last 20 years has extended its powerful synthetic methodology to make materials where size and shape are as important as structure. In other words, we have learned how to make nanoscale building blocks of different sizes and shapes, composition, and surface structure such as in the case of the "nano ink" developed by Nanosolar to make its CIGS panels.

The unique physicochemical features that only become accessible on the nanoscale, to achieve improved functionality and efficiency, are the optimal match of time and length scales for energy carrier transport and conversion, which can be achieved via nanoscale design of the photocatalytic site (Figure 1.9).

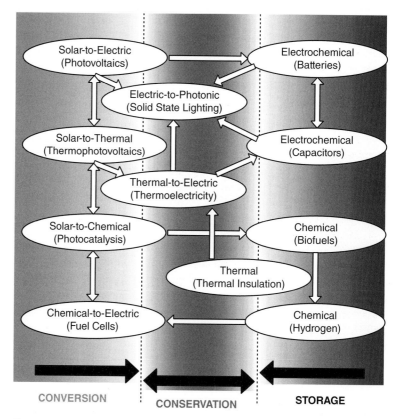

Figure 1.9
Portfolio of solar/thermal/electrochemical energy conversion, storage, and conservation technologies, and their interactions, that are the focus of the discussion. The electric grid is also shown in this figure as a network of connecting multiple elements (technology boxes), allowing them to act as a coherent whole. Reproduced from Reference [10], with permission.

Table 1.1 Characteristic length and time scales for energy carriers under ambient conditions. Reproduced from Reference [10], with permission.

	Wavelength (nm)	Mean free path (nm)	Relaxation time (ns)
Photons (solar/thermal radiation)			
In liquids/gases	100–30 000	>1000	>10^{-6}
In semiconductors	25–30 000	>10	10^{-7}–10^{-6}
In conductors/metals	–	0.1–10	10^{-10}–10^{-9}
Electrons			
In semiconductors/dielectrics	1–50	1–500	1–10×10^{-3}
In conductors/metals	0.1–1	1–10	10–100×10^{-6}
Phonons			
In semiconductors/dielectrics	0.5–10	1–500	10^{-3}–1
Molecules/ions			
In gas/plasma	10^{-2}–1	10^{3}–10^{7}	1–100
In liquid/electrolyte	–	0.1–1	10^{-3}
In solid/electrolyte	–	0.1–1	10^{-3}

An understanding of the general characteristics of fundamental energy carriers is important in appreciating the connections between nanotechnology and energy. Table 1.1 summarizes the characteristic length and time scales for energy carriers in liquids, gases, and solids. These scales define the space–time envelope within which, if accessible, the manipulation of matter should critically affect the energy carrier transport and conversion processes, thereby enabling drastic improvement in the performance of energy systems [10].

Because the length scales in Table 1.1 are generally of the order 1–100 nm, this size regime naturally falls into the domain of nanoscale science and engineering by virtue of the match between the length scales of the energy carriers and the materials that control their transport. Revolutionary improvements (i.e., an order of magnitude or greater enhancement when scaled to practical sizes) in the delivery of usable energy can be facilitated by advancing nanoscale design of both materials and associated energy conversion processes and systems.

For example, the present photovoltaic technologies rely on the quantum nature of light and semiconductors that are fundamentally limited by the band-gap energies. A revolutionary new approach suggested by Bailey in 1972 that revolves around the wave nature of light is now becoming a reality thanks to advances in nanochemistry [11]. The idea is simple: to use broadband rectifying (nano)antennas for direct conversion of solar radiation into electricity. The challenges in actually achieving the dream are many. The antenna concept relies on the fact that solar radiation is electromagnetic in nature. In other words, the waves are oscillating electric and magnetic fields propagating from the Sun to the Earth. In-coming light waves oscillate

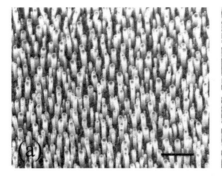
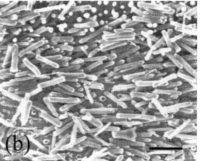

Figure 1.10
Aligned (a) and scratched (b) random arrays of MWCNTs are an array of dipole antennas. Each nanotube in such arrays is a metallic rod about 50 nm in diameter and about 200 to ca. 1000 nm long. Reproduced from rsc.org (Royal Society of Chemistry), with permission.

electrons in an antenna tuned to the wavelength of light – hence the devices have to be of the order of a few hundred nanometers. The broadband incoherent nature of the solar spectrum also requires a wide range of antenna sizes to match all the wavelengths and the need to arrange two directions of polarization.

These *rectennas* would not have the fundamental limitation of a semiconductor band-gap limiting their conversion efficiencies. How then can we manufacture rectennas with dimensions of the order of the wavelengths of solar radiation (mostly in the sub-micron range)? Using carbon nanotubes, a new form of highly conductive carbon accidently discovered in 1991.

Cheap and readily available multi-walled carbon nanotube (MWCNT) arrays grown on a metal substrate (Figure 1.10), in particular, behave as excellent optical rectennas, receiving and transmitting light at ultraviolet (UV), visible and infrared (IR) frequencies. In other words, these materials behave exactly like macroscopic antennas, namely they show a resonant response behavior as a function of the radiation wavelength. This arises from the condition that the induced current oscillations must fit into the antenna length.

Chemistry-enabled nanotechnology here resolves the biggest problem in achieving the ability to rectify electromagnetic waves at the high frequency range of visible and IR radiation. Although infrared rays create an alternating current in the nanoantenna, in fact, the frequency of the current switches back and forth ten thousand billion times a second, much too fast for electrical appliances, which operate on currents that oscillate only 60 times a second. Indeed, the *desired* length and diameter of the carbon nanotubes (CNTs) are achieved by accurate chemical control of the growth parameters. New nanochemistry strategies for the chemical synthesis of nanotubes developed in the early 2000s are therefore crucial to

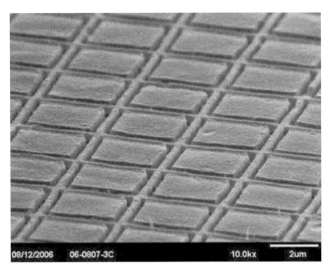

Figure 1.11
An array of loop nano-antennas imprinted on plastic and imaged with a scanning electron microscope. The deposited wire is roughly 200 nm thick. Photograph courtesy of Idaho National Laboratory.

manufacturing solar rectennas. The manufacturing process patented by the Idaho National Laboratory clearly demonstrates that nano-scale features can be produced on a larger scale [12].

The process stamps tiny square spirals of nanoantennas onto a sheet of plastic that is a flexible substrate. Each interlocking spiral nanoantenna is as wide as 1/25th the diameter of a human hair (Figure 1.11).[3] The nanoantennas absorb energy in the infrared part of the spectrum, just outside the range of what is visible to the eye. Since the sun radiates a lot of infrared energy, some of which is soaked up by the earth and later released as radiation for hours after sunset, nanoantennas can take in energy from both sunlight and the earth's heat, with higher efficiency than conventional solar cells. Double-sided panels could absorb a broad spectrum of energy from the sun during the day, while the other side might be designed to take in the narrow frequency of energy produced from the earth's radiated heat.

Present day light-to-electricity conversion efficiencies – measured under standard and *unrealistic* conditions – are in the 3–15% range, but in real applications the overall productivity of thin-film second-generation PV technologies is high. This, along with the lower price, renders the new PV technologies ready to provide cheap, clean electricity to both the 2 billion people who lack access to the grid and to energy eager companies and

3) See the Idaho National Laboratory Feature Story Archive at http://www.inl.gov/featurestories/2007-12-17.shtml.

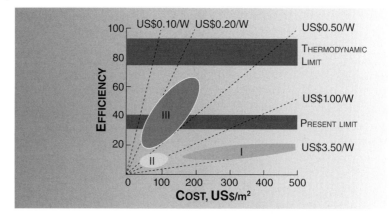

Figure 1.12
Efficiency and cost projections for first-, second- and third-generation photovoltaic technology (wafers, thin-films, and advanced thin-films, respectively). Reproduced from pv.unsw.edu.au, with permission.

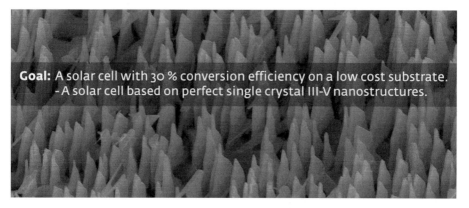

Figure 1.13
The objective of Sun Flake is clear: 30% efficient solar cells on all sorts of cheap substrates Reproduced from sunflake.dk, with permission.

families in the developed world facing the increasing cost of electricity generated with fossil fuels. In fact, much higher conversion efficiencies are to be achieved with the introduction of third-generation PV technologies (Figure 1.12).

The Danish company SunFlake (Figure 1.13) claims to be able to manufacture a low-cost cell that has an efficiency of about 30%–roughly double the efficiency of the average solar cell available today–through nanowire crystalline structures ("nanoflakes") of a semiconductor that can absorb nearly all light directed at them.

In conventional solar cells made of slabs of crystalline material, greater thickness means better light absorption, but it also means that it is more difficult for electrons to escape. This forced trade-off is overcome with *nanowires* [13]. Each nanowire is 10–100 nm wide and up to 5 μm long. Their length maximizes absorption, but their nanoscale width permits a much freer movement and collection of electrons. The light is absorbed essentially perpendicular to how electricity is collected and so the dilemma is overcome. The surface of the material consists of extremely tiny crystals of InAs, grown by molecular beam epitaxy (Figure 1.13), which provide a huge surface area in which sunlight can be caught.

A similar approach relies on nanowires containing multiple layers of group III–V materials, such as gallium arsenide, indium gallium phosphide, aluminum gallium arsenide, and gallium arsenide phosphide. It creates tandem or multi-junction solar cells that can absorb a greater range of the light spectrum than silicon [14] In conventional crystalline solar cells, group III–V materials have much higher efficiencies than silicon, but the great cost of these materials has limited their use. In addition to reducing costs by using less active material, this approach can cut the cost of the substrate that the nanowires are grown on, by using more plentiful and relatively cheaper silicon, glass, and polymer films for inexpensive roll-to-roll manufacturing.

1.5
Expanding the Solar Business

Two other new nanotechnology-based PV technologies are plastic and dye solar cells. Like CIGS-based cells, organic solar cells (OSCs) are particularly attractive because of their ease of processing, mechanical flexibility, and potential for low cost printing of large areas. Once higher efficiency is achieved such polymeric solar cells will rapidly find widespread application [15]. Indeed plastic solar cells are intrinsically cheap and easily manufactured, and being lightweight and flexible they will be rapidly integrated into existing buildings.

According to a set of design rules specifically derived for organic tandem cells, maximum efficiencies of 15% are expected for an optimized material couple [16].

Organic solar cells can be easily produced by roll-to-roll inkjet printing technology, as first demonstrated by Konarka Technologies (www.konarka.com) in early 2008 when the company, which specializes in organic photovoltaics, successfully carried out the first demonstration of manufacturing solar cells by inkjet printing. As in the case of the printed CIGS-based solar cells manufactured by Nanosolar, the continuous roll-to-roll printing technology provides easy and fast deposition of PV films over large area

Figure 1.14
Translucent solar cell modules. Wiring density gradually increases from left to right. This is to allow customers to select their favorite density. Reproduced from konarka.com, with permission.

substrates. Cleanliness during deposition is important. However, rather than encase the whole production line in a clean room to keep out dust, only the coating stations are sealed off. This lets the entire line reside in an ordinary factory environment. Modules made of Power Plastic® (Figure 1.14) coming off Konarka's roll-to-roll line today feature an energy conversion efficiency of 3% inside a building and 3–4% outside, because their high circuit resistance impacts more as current increases.

Invented in the early 1990s, dye-sensitized solar cells (DSCs) were first commercially used in 2003; the first modules based on this versatile hybrid (organic–inorganic) technology were used to fabricate one wall at Australia's new CSIRO Institute (Figure 1.15). As with plastic solar cells, DSCs have a low weight and low cost of production. However, their typical 7% efficiency in commercial modules is about twice that of polymeric modules; whereas their good performance in diffuse light conditions is a feature they have in common with inorganic thin-film solar modules. Finally, dye cells work well in a wide range of lighting conditions and orientation, as they are less sensitive to partial shadowing and low level illumination, making DSC-based modules particularly well suited for architectonic applications.

Beyond its low cost (titania is widely used in toothpastes, sunscreen, and white paint) and ease of production, the unique advantages over Si-based cells lie in their transparency (for power windows), easy bifacial configuration (advantage for diffuse light) and versatility (the color can be varied by selection of the dye, including invisible PV-cells based on near-IR sensitizers). By 2015, it is expected that companies will attain 10%-efficient modules that approach the criteria for solar module certification.

Figure 1.15
The first 4% efficient semitransparent orange windows consisting of DSC-based modules were used as elegant construction elements for the western façade of Australia's Energy CSIRO Institute in 2003. Photograph courtesy of Dyesol.

The Australian company Dyesol (www.dyesol.com) has pioneered the commercialization of DSC after obtaining a license from the inventors and has developed the technology in practically every aspect. The company has recently introduced a flexible, foldable, lightweight, and camouflaged solar panel for military applications that has been found to be superior to other PV technologies in maintaining voltage under a very wide range of light conditions – even in the dappled light under trees (Figure 1.16).

In general, DSCs are very tolerant to the effects of impurities because both light absorption and charge separation occur near the interface between two materials, which allows for roll-to-roll production, such as in the case of the G24 Innovations manufacturing process, transforming a roll of metal foil into a 45 kg half-mile of dye-sensitized thin film in less than three hours (see www.g24i.com).

This material is rugged, flexible, lightweight, and generates electricity even indoors and in low light conditions. In 2007, the US company G24 Innovations (www.g24i.com) started production in a factory based in the UK. Its modules are less than 1 mm thick and the original production capacity of 25 MW in 2007 is planned to be scaled up to 200 MW in a few years, following market response to initial commercialization. The first product is a solar charger series, working indoors and outdoors, compatible with mobile phones, laptops, audio players, and digital cameras. In Israel, the company 3GSolar (www.3gsolar.com) has developed inexpensive modules based on 15 × 15 cm dye cells, utilizing a low-cost method of depositing TiO_2 in a sponge-like array on top of flexible plastic sheets (Figure 1.17).

The company has developed a manufacturing line that costs 40% less than a silicon line per megawatt output, operating efficiently at 8 MW. This

18 1 *Capturing Sun's Energy*

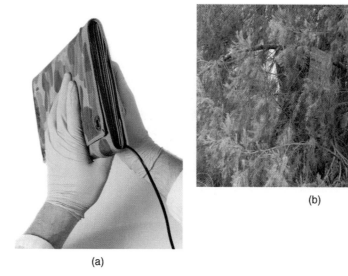

Figure 1.16
The flexible DSC-based solar module developed by Dyesol for the Australian Army camouflages itself in trees, where it provides constant voltage under a wide range of illumination levels Photograph courtesy of Dyesol.

Figure 1.17
Cameroon Minister of Science and Research, Madeleine Tchuinte, visiting 3GSolar headquarters and demonstration site in Jerusalem, September 25, 2008. Photograph courtesy of 3GSolar.

means that manufacturing can be put in place at 15% of the capital cost of a typical silicon photovoltaic line.

1.6
Solar Hydrogen from Water

Solar energy is inherently intermittent and must be stored to fulfill demand when irradiation is low or simply absent. Hydrogen (H_2), in contrast, is an excellent energy vector, suitable for storing solar energy, as clearly demonstrated by several recent commercial technologies of disruptive potential.

In Italy the building "HydroLAb" located in San Zeno, Arezzo, Tuscany (Figure 1.18) is completely autonomous and disconnected from all ordinary facility grids: water (rain water collection), sewage (bioremediation plants and close cycle), electricity (PV panels and fuel cells), gas (electrolysis hydrogen), phone and web (via radio) (www.lafabbricadelsole.it). Fully authorized by the local public safety and hygiene authorities in 2007, the building is the daily work place of about 20 knowledge workers in the renewable energy business.

A high efficiency membrane electrolyzer is used to produce renewable hydrogen that is then stored in ordinary storage systems or converted into

Figure 1.18
The HydroLAb in San Zeno, Arezzo, Italy. The system has a nominal power of 3.5 kWp and can produce 4200 kWh y^{-1} because of an estimated safety ratio of 1200 kWh kWp^{-1}, typical of the Arezzo latitude and microclimate. Reproduced from lafabbricadelsole.it, with permission.

metal hydrides. The stored energy can be recovered by mean of two 1 kW portable fuel cells.

A solar photovoltaic roof made of 20 conventional 175 W photovoltaic panels provides the electricity needed to run an electrolyzer that splits water into hydrogen and oxygen, which are then collected separately at the electrodes. When electricity is needed, both gases are supplied to the fuel cell, which generates electric power. For at least two years, the HydroLAb building has shown convincingly that a solar-powered economy is entirely possible from both an economic and environmental viewpoint using present-day technologies, and that solar-generated hydrogen will play a central role in tomorrow's energy infrastructure.

On the larger scale required to make solar-based H_2O splitting a practical technology in terms of quantity and cost, a simple and robust catalyst, with manufacturability on a large scale at competitive costs, must be developed. Awarded the EU Descartes Prize in 2006, the Hydrosol project led by Athanasios Konstandopoulos has indeed resulted in the development of a method of producing hydrogen from water-splitting using only the energy of the sun [17].

The process – an endothermic reaction that requires energy input – employs a multichannel ceramic honeycomb reactor, resembling the familiar catalytic converter of automobiles, coated with active water-splitting materials that are heated by concentrated solar radiation to the desired temperature. The reactor contains no moving parts and is constructed from special ceramic multi-channeled monoliths that absorb solar radiation, coated with active water-splitting nanomaterials capable of splitting water.

When water vapor passes through the reactor, the coating material splits the water molecule by "trapping" its oxygen and leaving in the effluent gas stream pure hydrogen. In the next step, the oxygen "trapping" material is regenerated with the help of solar power (i.e., the material releases the oxygen absorbed) by increasing the amount of solar heat absorbed by the reactor; hence a cyclic operation is established on a single, closed reactor/receiver system. Multi-cyclic solar thermochemical splitting of water was successfully demonstrated on a pilot solar reactor (Figure 1.19) achieving constant hydrogen production exclusively at the expense of solar energy.

The work has attracted interest from several international organizations, including the UN, which foresees huge potential for technological transfer to developing countries with high solar potential, thereby offering the prospect for the creation of new markets, as well as new energy sources. Once again, nanochemistry makes it possible as it enables the synthesis, through "aerosol" and combustion techniques, of the iron-oxide-based redox catalysts capable of producing pure H_2 from water in 80% yield; the catalysts take oxygen from water at reasonably low temperatures (800 °C) and are regenerated at temperatures below 1300 °C. Eventually, the integration of solar energy concentration systems with systems that can split water will have an immense impact on energy economics, as it is the most promising

Figure 1.19
In March 2008 a 100-kW reactor for producing hydrogen through water splitting using solar energy was put into commission at the Plataforma Solar in Almería as part of the Hydrosol project. The reactor is located inside the tower on the right. Reproduced from dlr.de, with permission.

route to providing affordable, renewable solar hydrogen with virtually zero CO_2 emissions.

Early in 2009 the Austrian companies Fronius, Bitter, and Frauscher successfully presented Riviera 600, the world's first electric boat powered by hydrogen fuel cells, on Lake Traunsee in Austria; the boat is featured on the cover of this book. The concept is that of self-contained energy supply provided by hydrogen obtained simply by photovoltaic electrolysis of water (www.zukunftsprojektwasserstoff.at). Extracted from water using photovoltaics and electrolysis, H_2 is oxidized in the fuel cell and the only emission is clean water (Figure 1.20), thereby completing a zero emission energy production cycle.

The boat – 6 m long, 2.2 m wide, and weighing 1400 kg – has a range of 80 km with a full hydrogen tank and has been awarded a safety certificate by Germany's TÜV. Its 4 kW continuous power electric motor has twice the range of conventional battery-powered boats. The 47% efficiency of the noise-free fuel cell engine should be compared to the 18–20% efficiency of a conventional (steel) internal combustion engine.

Its main economic advantage compared with conventional electric boats is the fact that no time has to be spent charging the batteries. For conventional electric boats, 6–8 h of charging gives just 4–6 h of use. The hydrogen-powered electric boat requires only the time that it takes to change the cartridge: 5 min. The boat's fueling system consists of a 20 kg cartridge that can be charged with up to 0.7 kg of hydrogen kept at 350 bar. Refueling is

Figure 1.20
The only emission of the Frauscher Riviera 600 hydrogen-powered boat is clean water; the H_2 is obtained cleanly by photovoltaic electrolysis of water. Reproduced from fronius.com, with permission.

carried out using a standard filler coupling, on the one hand, plus the simple exchange of an empty cartridge for a full one on the other hand (Figure 1.21).

The energy filling ("Clean Power") station makes use of PV modules integrated in a 250 m² flat roof, and further connected to an electrolytic cell. Even at Austria's cold latitudes the station can produce an annual yield of 823 kg of hydrogen, which is equivalent to 1100 cartridges with a 27 200 kW power content, that is, enough hydrogen to run to boat for 8000 km. Its installation is simple thanks to the "container construction" design and can be carried out simply and quickly at many different locations. The station consists of electricity power charger, hydrogen, and payment units (Figure 1.22).

Renewables expert Claudia Bettiol from Italy rightly argues that renewable energy will be a local, hi-tech business with a global impact. We add to this our view that it will increasingly make use of alternative forms of financing such as international organizations and Islamic finance. It is thus remarkable to see how this is in practice occurring. All three companies involved in the "Future Project Hydrogen" are based in Austria, and are

Figure 1.21
The 600 Riviera Frauscher boat can be refueled in 5 min using a standard 350 bar filler coupling plus the simple exchange of an empty cartridge for a full one. Reproduced from frauscherboats.com, with permission.

Figure 1.22
The Clean Power Station makes use of a 250 m² flat roof equipped with PV modules whose electric power output feeds an electrolyzer that splits water molecules. Reproduced from fronius.com, with permission.

very close to each other. Scientific and technical advice was provided by the Technical University of Graz, whereas the project was realized with support from the European Union regional programs and further funding from an Austrian region. The first 600 Riviera boat is commercialized at €150 000, with the first exemplars to be delivered to customers in early 2010.

References

1 International Energy Agency (2006) 2006 Key World Energy Statistics. http://www.iea.org/Textbase/nppdf/free/2006/Key2006.pdf.
2 IPCC (2007) Summary for policymakers, in *Climate Change 2007: Impacts, Adaptation and Vulnerability, Contribution of Working Group II to the Fourth Assessment Report of the Intergovernmental Panel on Climate Change* (eds. M.L. Parry, J.P. Palutikof, P.J. van der Linden, and C.E. Hanson), Cambridge University Press, Cambridge, UK, pp. 7–22.
3 Recent publications indicate that this temperature increase will likely occur at average atmospheric concentrations of CO_2 as low as 350 ppm. Hansen, J., Sato, M., Kharecha, P., Beerling, D., Berner, R., Asson-Delmotte, V., Pagani, M., Raymo, M., Royer, D.L., and Zachos, J.C. (2008) *Open Atmos. Sci. J.*, **2**, 217.
4 Nordhuas, T. and Shellenberger, M. (2007) *Break Through: From The Death of Environmentalism to the Politics of Possibility*, Houghton Mifflin, New York.
5 Market Buzz (2009) Complete solar industry statistics can be found at data available at: www.solarbuzz.co/Marketbuzz2009-intro.htm (accessed September 30, 2009).
6 *PV Perspectives* (October 2009), iSuppli (orp: El Segundo: CACUSA).
7 Hoffmann, W. (2004) *A Vision for PV Technology up to 2030 and Beyond – An Industry View*, European Photovoltaic Industry Association, Brussels.
8 Bjørn, L. (2001) *The Skeptical Environmentalist*, Cambridge University Press.
9 Fehrenbacher, K. (2007) 10 Questions for Nanosolar CEO Martin Roscheisen, Benchmark Capital, http://www.benchmark.com/news/sv/2007/07_30_2007a.php (accessed 30 July 2007).
10 Baxter, J., et al. (2009) Nanoscale design to enable the revolution in renewable energy. *Energy Environ. Sci.*, **2**, 559–588.
11 Bailey, R.L. (1972) *J. Eng. Power*, **94**, 73–77.
12 Ren, Z., Kempa, K., and Wang, Y. (2007) Solar cells using arrays of optical rectennas US Patent 20070240757.
13 Aagesen, M., Johnson, E., Sørensen, C.B., Mariager, S.O., Feidenhans, R., Spiecker, E., Nygård, J., and Lindelof, P.E. (2007) Molecular beam epitaxy growth of free-standing plane-parallel InAs nanoplates. *Nat. Nanotechnol.*, 761–764.
14 Chen, S.S., Fradin, C., Couteau, C., Weihs, G., and LaPierre, R.R. (2007) Self-directed growth of AlGaAs core-shell nanowires for visible light applications. *Nano Lett.*, **7**, 2584.
15 Brabec, C.J. (ed.) (2008) *Organic Photovoltaics*, Wiley-VCH Verlag GmbH, Weinheim.
16 Ameri, T., Dennler, G., Lungenschmied, C., and Brabec, C.J. (2009) Organic tandem solar cells: a review. *Energy Environ. Sci.*, **2**, 347.
17 Agrafiotis, C., and Konstandopoulos, A.G. (2005) Solar Hydrogen via Water Splitting in Advanced

Monolithic Reactors for Future Solar Power Plants, Technology Platform Operation Review Days, Brussels 8th–9th December 2005, https://www.hfpeurope.org/uploads/1105/1611/HYDROSOL-II_KONSTANDOPOULOS_TechDays05_051205_FINAL.pdf (accessed September 30, 2009).

18 A comprehensive, recent summary of the world energy portfolio and costs can be found in the UN-commissioned publication: UNDP, UNDESA and the World Energy Council (2004) World Energy Assessment Overview: 2004 Update. http://www.undp.org/energy/weaover2004.htm.

2
From Chemistry to Nanochemistry

2.1
Why Small is Different

For larger material aggregates, the deviation of properties from the bulk limit scales with the size of the aggregate. However, in most cases, as the size approaches the nanoscale regime, the dependence of material property on size becomes *non-scalable*; at this point, small is different in an essential way, with the physical and chemical properties becoming *emergent* in nature, that is, they can no longer be deduced from those known for larger sizes. This is the main physical concept at the basis of nanotechnology.

> When the non-scalable regime is approached, the physical size of the system becomes comparable to a phenomenon-dependent characteristic length.

The exemplars are many and include the photovoltaic effect due to nanocrystals described in Chapter 1, chemical catalysis, surface protection with nanocoatings, and much else. Modern studies in materials science focusing on systems with nanometer dimensions are thus generally aimed at elucidating evolution of the properties of materials from the molecular and cluster regimes to the so-called "bulk" phase. Very often, these studies make use of computer-based simulations, which serve as a powerful predictive tool supplementing and complementing laboratory experiments, allowing researchers to gain insight into the microscopic physical and chemical origins of the behavior of materials [1].

For example, it is its well-known inertness that provides gold with its striking "noble" metal value. Yet, when shrunk to nanoparticle size, gold turns into a powerful reactive material acting as a recyclable catalyst (Figure 2.1). Briefly, small metal clusters exhibit unique properties that originate from the highly reduced dimensions of the individual metal aggregates [2]. In particular, investigations on size-selected small gold clusters, Au_n ($20 \geq n \geq 2$), have revealed that gold octamer clusters (Au_8)

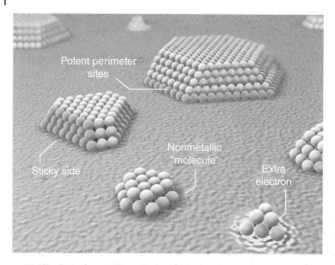

Figure 2.1
Chemists have proposed several possible mechanisms to explain why ordinarily inert gold turns into a powerful catalyst when shrunk to nanoparticle size. Reproduced from Reference [2], with permission.

bound to a magnesia surface can oxidize CO into CO_2 at temperatures as low as 140 K. In *juxtaposition* with laboratory experiments, computer-based simulations led investigators to conclude that the charging of the metal cluster, caused by partial electron transfer from the substrate into the deposited cluster, underlies the catalytic activity of the small gold clusters (Au_n, $8 \leq n \leq 20$).

In 1999 these investigations predicted that (i) oxygen vacancies on the metal-oxide support surface play the role of "active-sites;" (ii) these sites serve to strongly anchor the deposited clusters, thus inhibiting their migration and coalescence; and (iii) these active sites control the charge-state of the gold clusters, thus promoting the activation of adsorbed reactant molecules. Theoretical chemists thus suggested that it is crucial to use imperfect crystals of magnesia as it is the surface of the metal oxide that is responsible for the enhanced activity of the adsorbed gold cluster. The results have indeed been observed experimentally and several new commercial gold catalysts have entered the market since 2007 (www.goldinnovationsblog.com). However, the reader should not be mislead into thinking that this example of the potential of computers in materials science shows that nanotechnology is a theoretical construct mainly developed by computer science. In juxtaposition with computer studies, nanotechnology advances of practical interest, such as with gold catalysts, have been made possible by advances in chemistry and, in particular, in nanochemistry.

2.2
Nanochemistry, the Chemical Approach to Nanotechnology

The essence of nanotechnology is the ability to work on the nm-length scale to generate *larger* structures with fundamentally *new* molecular organization [3]. In this sense, one might argue that nanotechnology is the marriage of engineering and chemistry to create nanoscale objects that exhibit novel materials properties, largely as a consequence of their finite small size. Yet, whereas both the nano-engineering and nanophysics approach towards this goal tended to operate from the bulk "down," the chemical approach to nanotechnology has been from the atom "up:" Building and organizing nanoscale objects under mild and controlled conditions "one atom at a time" instead of "manipulating" the bulk. In 1992, the father of the concept, the British scientist Geoffrey Ozin (Figure 2.2), published what can be considered the nanochemistry manifesto, calling for a chemical approach to nanotechnology. Therein Ozin concluded that the approach suggested should in principle "provide a reproducible method of producing materials that are perfect in size and shape down to the atoms" [4].

Ozin's insight happened to be dramatically fertile. In the following two decades the first examples of successful nanotechnology innovations of practical impact–including all those discussed in this book–have been actually based on nanochemistry, that is, the utilization of synthetic

Figure 2.2
Geoff Ozin, a British scholar professor at the University of Toronto, developed the concept of nanochemistry. Photograph courtesy of Geoff Ozin.

chemistry to make nanostructured materials with special properties. We can even argue that this evolution is another surprising example of *historical hysteresis*, in which the revival of interest in areas of research such as materials morphogenesis described in the classic *Of Growth and Form* of D'Arcy Thomson (1917) has led to dramatic advances in creating one of the most active and interesting fields in chemistry and materials science research today, one that we had come to consider as completely closed.

In the typical pragmatic view of chemists, Ozin explained that nanotechnology had to do with making *useful materials* and devices with unique properties through assembly of atoms or molecules on a scale between that of the individual building blocks and the bulk material. At this level, quantum effects can be significant, and innovative ways of carrying out chemical reactions also become possible. Chemists then would have developed the concepts and methods involved in synthesizing nanoscale building blocks with controlled size, shape, structure and composition. In brief, nanochemistry (Figure 2.3), as opposed to nanophysics, emphasized the synthesis rather than the engineering aspects of preparing little pieces of matter with nanometer sizes in one, two or three dimensions. Put another way:

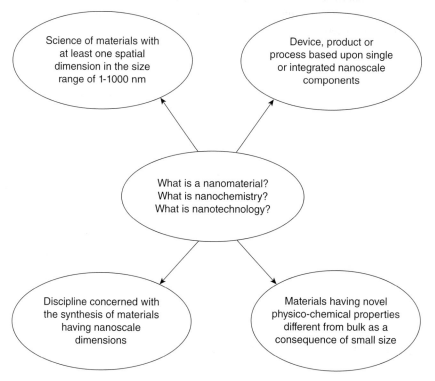

Figure 2.3
Nanomaterial, nanochemistry, and nanotechnology. Reproduced with permission of G. A. Ozin.

> "In the *Solid State 20th Century*, where trial-and-error solid-state synthesis produced unique structures, structures produced new physics and physics produced novel devices, it has been the fields of materials science and engineering that have propelled materials technology to the spectacular heights that society accepts and enjoys today."
>
> "As we enter the *Materials 21st Century*, where designed chemical synthesis creates original molecules, molecules produce new materials, materials produce novel structures and structures produce unique function and utility, it is the insatiable demand for new materials in emerging fields like nanotechnology and biotechnology that is now driving the fields of materials science and engineering" [5].

In satisfying this "insatiable" demand chemists, the inventors and makers of new useful substances, again play the same central role occupied in the development of the iconic industry of the twentieth century, the chemical industry. Today, the chemical industry is the world's largest in terms of size and revenues; and is certainly among the most relevant as without its products modern society would simply not exist.

As chemistry turns into nanochemistry, therefore, an insight into chemistry's conceptual foundations is useful to nanotechnology and nanochemistry practitioners, be they researchers, managers, or educators. A powerful, autonomous chemical methodology in fact exists that is based on the visualization and association of chemical "building blocks." It is such methodology that originates the cornucopia of ever new, artificial substances that largely benefit society, the economy, and – yes – the environment. In brief:

> "What we are learning from a decade and a half of nanochemistry research is that nanochemistry is a 'provider' of a myriad of nanoscale building blocks, an 'enabler' of nanotechnology, and a contributing 'founder' of future bottom-up nanofabrication, the success of which will be contingent upon being able to synthesize and assemble single size and shape building blocks with controlled surfaces into functional nanostructures and integrate them into useful and defect-tolerant nanosystems" [6].

2.3
An Insight into Chemical Methodology

At the core of chemical methodology lies a powerful approach based on mental visualization and association of chemical *models* for substances. These models can be molecular structures. But they can also be synthetic "building blocks" of different size and shape, as the recent development of nanochemistry clearly shows.

Asked why two electrically negative particles called electrons should actually bind two atoms together, our student remained puzzled. "Both negative? I never thought about that." Probably, our student is not alone. After all, chemists are experimental scientists who are supposed to spend most of their time mixing smelly stuff in the laboratory without devoting much attention to these conceptual details. And yet, is this true? What is the conceptual difference between what a layman thinks the practice of chemistry is and what it really is?

The eminent physicist Paul Dirac, for instance, had little doubt, claiming in 1929 that:

> "The fundamental laws necessary for the mathematical treatment of a large part of physics and the whole of chemistry are thus completely known, and the difficulty lies only in the fact that application of these laws leads to equations that are too complex to be solved" [7].

According to this view, chemistry should be considered as a well understood branch of physics. However, how far all this is from the practice of the chemical enterprise does not seem to disturb many contemporary chemists.

The pillars of chemistry are two, namely *analysis* and *synthesis*. Chemical analysis enables us to find out which elements – operationally defined by in 1789 by Lavoisier (Figure 2.4) as "the furthest stage to which analysis can reach" – compose a substance; chemical synthesis allows chemists to create new substances or to devise new routes to compounds created by Nature. The complexity of the substances that chemists can analyze and synthesize, as well as the ease of either process, has of course increased enormously over the past century thanks to the progress made both in theory and in the available experimental tools: a wealth of specialized reagents to carry out specific transformations; chromatography to isolate and analyze substances of interest; spectroscopy and other physical methods to rapidly elucidate the structures of unknown compounds.

Yet one of the great scientific revolutions of the twentieth century [8] occurred in the late 1920s when chemists developed a *theory* of chemical reactions based on the rearrangement of electron pairs as bonds are made and broken, opening the route to designing synthetic routes and ultimately putting organic synthesis on a logical basis.

From a chemist's viewpoint, the concept of molecular structure is a functional model that is essential for predicting the outcome of chemical reactions *not yet carried out*. What chemists really do is to create mental images of the substances they wish to create and then manipulate these shapes in a rational manner to verify if they could fit to afford the desired

Figure 2.4
Chemistry stands on two pillars: analysis and synthesis; and we still make use of the element concept developed by Lavoisier, shown here, at the end of the eighteenth century. Reproduced from Wikipedia.org, with permission.

substance. Kekulé's speech at a dinner commemorating his "discovery" of benzene's structure in 1865 (Figure 2.5) renders this process vividly:

> "The atoms were gamboling before my eyes ... My mental eye, rendered more acute by repeated vision of this kind, could not distinguish larger structures, of manifold conformation; long rows, sometimes more closely fitted together; all twining and twisting in snakelike motion. But look! What was that? One of the snakes had seized hold of its own tail, and the form whirled mockingly before my eyes" [9].

Therefore, the molecular structure is not, in and of itself, a manifestation of the notion that chemistry is subordinate to physics, but rather a powerful *model* chemists can use.

To show how fertile this visualization approach continues to be today let us take, for example, a chemical reaction.

The scheme in Equation 2.1 says that the reaction between an amine and a ketone *generally* occurs with the elimination of water to afford a valuable substance called an amide:

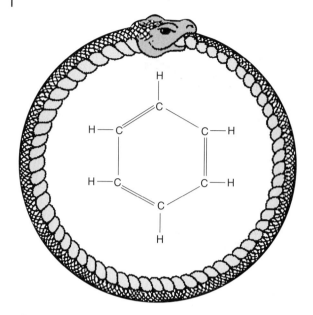

Figure 2.5
The structure of benzene revealed as a mental image of a snake to Kekulé. Reproduced from Reference [9], with permission.

$$R-NH_2 + R'R'-CO \rightarrow R-N=C-R'R'' + H_2O \qquad (2.1)$$

This general scheme is a representation of an experimental process. Chemists then infer combinatorial rules for the structures (models) representing substances. Under certain experimental conditions (which, in chemistry, are crucially important), pure substances behave according to these rules. In other words, such association and dissociation norms are based on repeated observation, *abstraction*, and *generalization*.

A mechanistic viewpoint that makes use of ideas such as molecular structure, valence, electrons, and so on is then applied to justify observations. But, certainly, no physicist would find this intellectual process anywhere close to what he/she actually does to interpret the results of physical experiments. Yet, chemists using their unique methodology have been able to synthesize a cornucopia of incredibly useful substances. This is what gives chemists their power and scientific importance and not the "self-imposed tyranny" [10] of the Schrödinger equation for the N-electron Ψ wavefunction (Equation 2.2):

$$H\Psi = E\Psi \qquad (2.2)$$

$$\Psi(1, 2, \ldots, N) = \varphi(1)\varphi(2)\ldots\varphi(N) \qquad (2.3)$$

In a modern plural view the molecular structure becomes a model for the substance, namely a *representation* of it or – even better said – "a symbolic

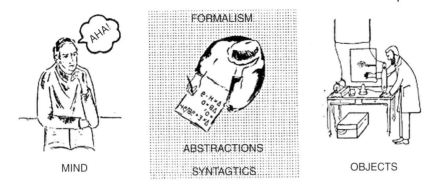

Figure 2.6
Chemists create mental images of the behavior of matter. They then go the laboratory to create their objects: the myriad substances so-produced benefiting society at large. Reproduced from Reference [19], with permission.

transformation of reality, at once a model and a theoretical construct" [11]. The chemist's mind proceeds by creating an image (Figure 2.6) of how atoms, molecules, and other matter building blocks will actually behave, and then tries to mentally "push" them along the desired route.

Put another way, what chemists actually *do* when practicing their science is to mentally play with these representations *before* entering the laboratory. In this sense, learning the practice of chemistry is analogous to learning a language. In each language, there are rules to combine the elementary units (words), which in their turn represent objects and ideas. The result of the combination is a meaningful language that enables people to communicate. In chemical practice, the units are the chemist's building blocks and the outcome is a new substance whose structure and functions are to be discovered. Chemistry constitutes a *language*:

> "A chemical formula is like a word. It purports to identify, to single out the chemical species it stands for ... but it remains a long, long way from the molecular scale to the macroscopic world of the senses. We still have to represent molecules ... And we tend to represent atoms *as if* they were normal objects in our everyday experience" [7].

Following the introduction of advanced physical instrumentation in the 1950s, chemists became able to determine the structure of the substances created in their experiments at an unprecedented rate, and thus check the match between the visualization process mentioned above and results of synthetic experimentation. This, in turn, further accelerated the creation of new heuristic combinatory rules offering increasing control over chemical reaction paths at such a level that the concept of molecular design emerged as a realistic objective.

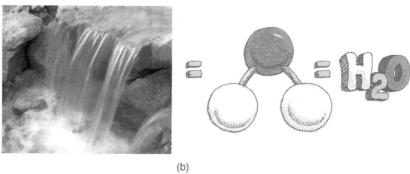

Figure 2.7
A representation is just a representation. As Magritte reminds us that a drawing of a pipe is not a pipe (a), so a molecular structure is not a chemical substance (b).

However, our scientific theories do not refer directly to *reality* but to a model [12]. A model in its turn is an abstract entity that selects from reality the variables that we consider relevant for our purposes. We make hypotheses about the unobservable objects, properties, and structure of the system under consideration and then we access real systems – that is, chemical substances – through our models. In other words, our direct access to real systems is invariably mediated by our ideas. This is what is meant by the idea of "representation" [13].

A model is a particular kind of representation; and exactly like a representation of an object is not the object (Figure 2.7), a model of a system is *not* the system.

In chemistry, too, the reductionist argument that there has to be one logical way to do things to be counted as adequate or successful is flawed. No ontological reduction of chemistry to the realm of physics is possible. The chemical world is interconnected to the world of physics by nonreductive links that permit the existence of objective relationships between both worlds, but preserve the ontological autonomy of each one [12].

As the historical development of nanotechnology has shown, the richness of the chemical methodology–bottom-up, from the atoms to the nanoscale objects–has led to the creation of a multiplicity of nanostructured materials that are finally providing the long-awaited benefits invoked by Feynman in his celebrated "There is plenty of room at the bottom" lecture (Figure 2.8) at the 1959 meeting of the American Physical Society:

> "A chemist comes to us and says, 'Look, I want a molecule that has the atoms arranged thus and so; make me that molecule.' The chemist does a mysterious thing when he wants to make a molecule. He sees that it has got that ring, so he mixes this and that, and he shakes it, and he fiddles around. And, at the end of a difficult process, he usually does succeed in synthesizing what he wants. By the time I get my devices working, so that we can do it by physics, he will have figured out how to synthesize absolutely anything, so that this will really be useless."

> "But it is interesting that it would be, in principle, possible (I think) for a physicist to synthesize any chemical substance that the chemist writes down. Give the orders and the physicist synthesizes it. How? Put the atoms down where the chemist says, and so you make the substance. The problems of chemistry and biology can be greatly helped if our ability to see what we are doing, and to do things on an atomic level, is ultimately developed–a development which I think cannot be avoided" [14].

In the subsequent 50 years, single atom "manipulation" of the type invoked above remained elusive; and the "bottom-up" approach first conceived by Richard Feynman was actually accomplished using chemistry. The photovoltaic "nano-ink" of copper-based semiconductor CIGS that is making cheap solar electricity a reality (Figure 2.9) is an explicit example of the potential of this approach. Embedded in a solar cell, 1 kg of CIGS produces five-times as much electricity as 1 kg of enriched uranium embedded in a nuclear power plant. Put another way, 1 kg of CIGS is equivalent to 5 kg of enriched uranium in terms of the energy the materials deliver as solar and nuclear power, respectively.

Other avenues, based on a "manipulation" approach, instead ended in disquisition about "nanorobots" and "nanomachines" and similar intellectual bottlenecks.

Remarkably, scholars studying the historical development of nanotechnology have clearly shown how Feynman's stature as an iconic figure in twentieth-century science was deliberately used to help advocates of nanotechnology by providing a valuable intellectual link to the past and giving nanotechnology a packaged history with an early start date in December 1959; whereas Feynman's actual role in catalyzing nanotechnology research

There's plenty of room at the bottom, says noted scientist as he reveals—

How to Build an Automobile

Exploring the fantastic possibilities of the very small should pay off handsomely— and provide a lot of fun, too

By Richard P. Feynman
*Professor of Theoretical Physics,
California Institute of Technology*

At 42, Richard Phillips Feynman, Ph.D., enjoys world renown as a theoretical physicist, local fame as a "marvelous" performer on the bongo drums, and campus admiration as a man with a pixyish humor that turns a lecture on quantum electrodynamics into a ball. You'll see why when you read his impassioned and witty plea to think small.

This tall, slim, dark-haired scholar helped importantly in developing the atomic bomb and watched its first test explosion. In 1954 he won the $15,000 Albert Einstein Award, one of the nation's highest scientific honors.

He is capable both of exuberant fellowship and of rather stern withdrawal, especially when pondering intricate problems. Even his heavy thinking has a light touch, however. In deepest thought, while pacing the floor, he slowly flips a silver dollar back and forth across the fingers of his right hand by carefully controlled movements of the knuckles. It's no easy trick even when you have nothing else to think about.

Born in New York City in 1918, he graduated from MIT in 1939 and got his Ph.D. at Princeton in 1942. He was a member of the Laboratory of Nuclear Studies at Cornell from 1945 to 1950. In 1950, he began his present job as professor of theoretical physics at Caltech.

Dr. Feynman loves music, children, camping in the wilds, and unpremeditated jaunts to faraway places. He boned up on Portuguese to become a visiting lecturer for two seasons in Brazil, and learned Spanish under forced draft to go to Peru and poke around Inca ruins.

The accompanying article is condensed from a speech (addressed to an American Physical Society meeting, not the Pasadena Rotary luncheon). The full transcript appeared in "Engineering and Science Magazine," published at the California Institute of Technology.

PEOPLE tell me about miniaturization, about electric motors the size of the nail on your small finger. There is a device on the market by which you can write the Lord's Prayer on the head of a pin. But that's nothing. That's the most primitive, halting step.

Why not write the entire 24 volumes of the "Encyclopaedia Britannica" on the head of a pin?

Let's see what would be involved. The head of a pin is a sixteenth of an inch across. If you magnify it 25,000 diameters, the area of the head of the pin is equal to the area of all pages of the encyclopedia. All it is necessary to do is to reduce the writing in the encyclopedia 25,000 times. Is that possible? One of the little dots on the fine halftone reproductions in the encyclopedia, when you demagnify it by 25,000 times, still would contain in its area 1,000 atoms. So, each dot can easily be adjusted in size as required, and there is no question that there is enough room on the head of a pin to put all of the "Encyclopaedia Britannica."

IMAGINE that it is written in raised letters of metal that are 1/25,000 ordinary size. How would we read it?

We would press the metal into plastic and make a mold; peel the plastic off very carefully; evaporate silica into the plastic to get a very thin film; then shadow it by evaporating gold at an angle against the silica so that all the little letters appear clearly; dissolve the plastic away from the silica film; and then look through it with an electron microscope.

Figure 2.8
Frontispiece of the November 1960 issue of the magazine *Popular Science*, a published account of Richard Feynman's celebrated talk on "the fantastic possibilities of the very small." Reproduced from blog.modernmechanix.com, with permission.

Figure 2.9
1 kg Uranium = 5 kg CIGS. Yet, the uranium is burned and then stored in a nuclear waste facility; the CIGS material produces power for at least the warranty period of the solar cell product, after which it can be recycled and reused an indefinite number of times. Reproduced from Nanosolar.com, with permission.

was limited, based on recollections from many of the people active in the nascent field in the 1980s and 1990s. Cultural anthropologist Chris Toumey (Figure 2.10), in particular, found that the published versions of Feynman's talk had a negligible influence until 1991 [15].

Subsequently, interest in "Plenty of Room" in the scientific literature greatly increased when the term "nanotechnology" gained serious attention following concomitant launch of the journal *Nanotechnology* in 1989; publication of the famous Eigler–Schweizer experiment, precisely manipulating 35 xenon atoms, in *Nature* in April 1990; and *Science*'s special issue on nanotechnology in 1991.

2.4
Making Nanomaterials

Nanochemistry is neither a complicated procedure performed on single molecules nor a purely academic topic where applications are beyond the horizon of our thinking. Instead, nanochemistry *expands* the reach of the chemical approach, extending chemical methodology to materials synthesis. The result is a physicochemical handling of matter that relies on secondary valences, the adjustment of geometrical shape and interface energies, and self-organization, rather than classical synthesis [16]. From a scientific viewpoint, the resulting "bottom-up" approach is a new way of

Figure 2.10
Published versions of Feynman's celebrated talk "There is plenty of room at the bottom" had a negligible influence until 1991, as cultural anthropologist Chris Toumey found out. Reproduced from cas.sc.edu, with permission.

thinking about the structure–activity relationships governing the behavior of functional materials.

Chemists, in practice, create new nanomaterials through the utilization of planned synthetic chemistry to produce molecular building blocks for purposeful construction of nanomaterials. In this sense, therefore, nanochemistry bridges the worlds of molecules connected by molecular bonds and the chemical engineering of micron-sized structures, such as lithography, chemical vapor deposition, or coating techniques.

Such nanoscale building blocks of different size and shape, composition, and surface structure can be useful in their own right, or in a self-assembled structure [17]. In general, for nanomaterials size and shape are *as important* as structure and composition. Or, as put it by an officer of the US Environmental protection agency:

Figure 2.11
"Functional randomness" has been invoked by Israeli chemist David Avnir as chemists are learning to construct complicated systems "by allowing disorder and correlations to compete and to come to terms with each other through an optimal solution." Photograph courtesy of the author.

> "We're in a scientific revolution like the one defined by Thomas Kuhn. The paradigm that we have shifted to is that properties change with size alone, not just composition. New tools and approaches are necessary to show that the phenomena were real. We have never actually believed it before" [18].

Defects in the assembly structure of these materials become central as it is imperfection that provides them with interesting properties and ultimately with function. Disordered sol-gel glasses doped with organic molecules are a celebrated example of an enormously large class of materials in which several different functions are dictated by imperfect and tunable geometry.

"In practice," says Israeli chemist David Avnir (Figure 2.11):

> "we are learning to master functional randomness, namely the teaching of Nature that is much more efficient to construct complicated systems by allowing disorder and correlations to compete and to come to terms with each other through an optimal solution."

Materials self-assembly, going into more detail, is the heart of nanochemistry as it has introduced an entirely new way of thinking about *how to make materials*. In a self-organizing system of materials a particular architecture

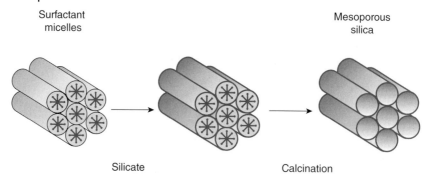

Figure 2.12
Materials synthesis by self-assembly is easily carried out by growing, for example, the material around the structure of a micelle in solution. Here the hexagonal arrays of cylindrical micelles formed in the solution dictate the growth of hexagonal silicate cylinders: periodic mesoporous organosilica.

forms spontaneously with a structural design that is determined by the size and shape of the individual nanocomponents and by the map of bonding forces between them. Such spontaneous organization of building blocks into assemblages that are unconstrained by scale is due to molecular forces that operate at length scales beyond the molecular, between the building blocks and over different scales.

For example, Figure 2.12 shows a schematic drawing of the liquid-crystal templating mechanism. Hexagonal arrays of cylindrical micelles are formed in the solution, and silicate species occupy the spaces between the cylinders. A new form of "mesoporous" silicate structures are thus obtained by simply burning out the organic templates. Such mesoporous silica—which has a very large surface area and shows an organized, periodic structure despite the amorphous character of the silica walls (see above the concept of functional randomness)—has found great utility as catalysts and adsorbents because of the regular arrays of uniform pore channels.

A brief analysis the historic development of materials chemistry starting, for instance, from Harting's work with biomineral formation (1873) through to the classic *Of Growth and Form* of D'Arcy Thomson (1917) shows how the effort to apply physico-geometrical principles to explain morphogenesis in the study of natural materials has been a constant driving force of scientific thought. Both modern materials chemistry and nanochemistry-enabled nanotechnology are a continuation of this intellectual trend. It is sufficient, for that matter, to recognize the effect of the discovery that organized organic molecules such as micelles in soap actually direct the growth of inorganic oxides in solution. This single finding brought about a revolution in the preparation of new porous materials: Dick Barrer first shows how quaternary aklylammonium cations dictate assembly of microporous alumino/silicates; Charlie Kresge explains how to extend the length scale well beyond molecular scale; and Edith Flanigen demonstrates how to go beyond aluminosilicates. Today, the artificial "zeolites" materials originated by these studies are among the most valued and best performing catalysts employed in the chemical industry.

References

1 Landman, U. (2005) Materials by numbers: computations as tools of discovery. *Proc. Natl. Acad. Sci. U.S.A.*, **102**, 6671.
2 Cho, A. (2003) Connecting the dots to custom catalysts. *Science*, **299**, 1684.
3 Antonietti, M., Landfester, K., and Mastai, Y. (2001) The vision of "Nanochemistry" or: is there a promise for specific chemical reactions in nano-restricted environments? *Isr. J. Chem.*, **41**, 1.
4 Ozin, G.A. (1992) Nanochemistry: synthesis in diminishing dimensions. *Adv. Mater.*, **4**, 612.
5 Ozin, G.A. (2001) Towards a Molecular Nanotechnology, Canadian Chemical News (November/December), 21–25. Available at http://www.chem.toronto.edu/staff/GAO/flashed/Geoff_files/nanotechnology1.pdf.
6 Ozin, G.A. and Cademartiri, L. (2009) Nanochemistry: what is next? *Small*, **5**, 1240–1244.
7 Dirac, P.A.M. (1929) Quantum mechanics of many-electron systems. *Proc. R. Soc. London A*, **123**, 714.
8 Corey, E.J. and Cheng, X.-M. (1995) *The Logic of Chemical Synthesis*, John Wiley & Sons, Inc., New York.
9 Cited in Weisberg, R. (1992) *Creativity, Beyond the Myth of Genius*, W. H. Freeman, New York.
10 Ogilvie, J.F. (1990) The nature of the chemical bond–1990: there are no such things as orbitals. *J. Chem. Educ.*, **67**, 280.
11 Hoffmann, R. and Laszlo, P. (1991) Representation in chemistry. *Angew. Chem. Int. Ed.*, **30**, 1.
12 Lombardi, O. and Labarca, M. (2005) The ontological autonomy of the chemical world. *Found. Chem.*, **7**, 125.
13 Hesse, M. (1966) *Models and Analogies in Science*, University of Notre Dame Press, Notre Dame.
14 Feynman, R.P. (1960) There's plenty of room at the bottom. An invitation to enter a new field of physics. 13th meeting of the American Physical Society, California Institute of Technology, December 29, 1959. *Eng. Sci.*, **23** (February issue), 22–36.
15 Toumey, C. (2008) Reading Feynman into nanotechnology: a text for a new science. *Techné*, **13**, 133.
16 Whitesides, G.M., Mathias, J.P., and Seto, C.T. (1991) *Science*, **254**, 1312.
17 Ozin, G., Arsenault, A., and Cademartiri, L. (2009) *Nanochemistry: A Chemical Approach to Nanomaterials*, RSC Publishing, Cambridge.
18 Cited in: Lubick, N. (2009) Promising green nanomaterials. *Environ. Sci. Technol.*, **43** (5), 1247.
19 Primas, H. (1983) *Chemistry, Quantum Mechanics and Reductionism*, Springer, Berlin.

3
Storing and Supplying Clean Energy

3.1
Ending the Era of the Internal Combustion Engine

When American troops entered Munich in March 1945, one of the first companies to be seized was Varta, the large battery manufacturer. The company was the manufacturer of batteries and accumulators for the Wehrmacht submarines and for the Luftwaffe during World War II [1]. It is batteries that enable storage and thus the *portability* of electric power; and since electricity is the most useful form of energy available to power technology devices, be they radios or noise-less submarine engines, batteries remain a key technology in virtually every industry. What, then, happened to the electric car (Figure 3.1)?

The obsolescence of battery technology can be immediately grasped by the short duration of the batteries of your Blackberry or of your iPhone (Figure 3.2), which makes users complain on web forums and blogs.[1] On a larger scale, among the main culprits in the downfall of the electric car in the USA was the limited range (60–70 miles) and reliability of the first electric car made available for lease in Southern California in 1991 (Figure 3.3).[2]

However, put simply, the combustion of hydrocarbon fuels in the internal combustion engine (ICE) for land, sea, and air transportation is the largest source of global warming pollution in the world – responsible for 30% of our annual CO_2 emissions [2]. Therefore, having to establish priorities in our efforts to stop global warming, the replacement the 150-year-old ICE with new, zero carbon emission vehicles is the single most important

---

1) See, for example, the post "Extend battery life of your iPhone 3G with tips from Apple" at: http://www.bloghash.com/2008/07/extend-battery-life-of-your-iphone-3g-with-tips-from-apple/.
2) This was identified in the 2006 documentary film on the history of the electric car (Figure 3.1). The film explored the creation, limited commercialization, and subsequent destruction of the battery electric vehicle in the United States in the 1990s (after California's Government passed the Zero Emission Vehicle mandate in 1990).

Nano-Age: How Nanotechnology Changes our Future. Mario Pagliaro
© 2010 WILEY-VCH Verlag GmbH & Co. KGaA, Weinheim
ISBN: 978-3-527-32676-1

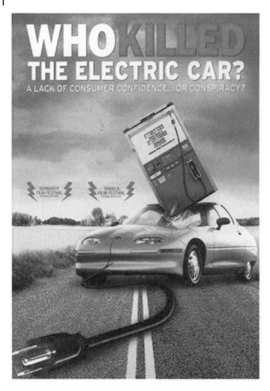

Figure 3.1
Image from a 2006 documentary film on the failure of the electric car in the USA. Image courtesy of Papercut Films, Copyright 2007.

Figure 3.2
Despite an enhanced battery life from 5 to 8 hours following the launch of the blockbuster iPhone in 2007, the lifetime of the battery remains too short – users of this and of other smartphone devices complain on the web. Photograph courtesy of Apple Inc.

Figure 3.3
General Motors launched the EV1, a battery electric vehicle, in the early 1990s. A few years later the car was retired from the market. Reproduced from Rmhermen/ wikipedia.org, with permission.

aim. This means that we need to develop efficient and affordable electric vehicles ultimately powered by the sun, either through photovoltaic electricity to recharge batteries or fueled by solar-generated hydrogen.

A new type of lithium-ion battery, which features an extremely long lifespan, may solve the problem. However, before entering the technology discussion we should lose a little of the naiveté that features in the arguments of too many scientists. The 150-year-old ICE is the basic infrastructure through which oil companies – the world's largest and most powerful companies – have established their global dominant position. Current efforts to replace the combustion engine with oil-free alternatives constitutes the single biggest threat these companies have ever faced. Frederic Beigberder, a former advertising star, has clearly explained the concept in an autobiographic look at the (top) advertising business:

> "We all knew that companies were keeping undisclosed many technologies that could threat their business; from ever-lasting tires to electric cars ..." [3].

This understanding is relevant, for example, in realizing *why* technologies that were publicly demonstrated some 30 years ago have been left dormant for decades. Readers in their forties, for example, are likely to have read in their country's press about "prototype" electric cars since their youth, being told that they would be too expensive; or they may recall the commercial failure of subsidized projects (Figure 3.1).

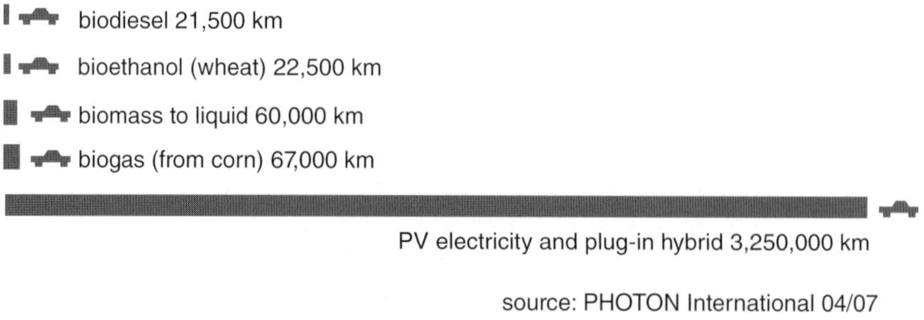

biodiesel 21,500 km

bioethanol (wheat) 22,500 km

biomass to liquid 60,000 km

biogas (from corn) 67,000 km

PV electricity and plug-in hybrid 3,250,000 km

source: PHOTON International 04/07

Figure 3.4
How far a car can drive on different forms of energy, each produced from 100 × 100 m of land. Reproduced from *Photon International* issue 4/2007, with permission.

The sustainability crisis made up of a mix of volatile oil price, global warming, and chronic air pollution is rapidly changing all that. And nanotechnology is the science behind the much needed new batteries and fuel cells required to change this state of planned obsolescence. Electric cars, for instance, are about four times more energy efficient than fuel based cars (Figure 3.4).

This efficiency arises because fuel engines mostly create heat and thus waste most of the energy units available. Biofuel plants are not efficient solar energy harvesters relative to semiconductor based solar electricity, and the result is the huge difference shown in Figure 3.4. In other words, while presently cars are baking in the sun all day, future all-electric cars will charge up while idling under a solar carport, as happens with the new Toyota Prius hybrid vehicle in which a solar module powers the air conditioner (Figure 3.5).

Figure 3.5
Using solar panels to power the air conditioning system, the solar version of the popular Toyota Prius hybrid vehicle will be available by the end of 2009. Reproduced from toyota.com, with permission.

3.2
Nanotechnology-Based Batteries

True success in the transportation market for electric vehicles depends upon a significant improvement in battery technology. Safety, recharge time, power delivery, extreme temperature performance, environmental friendliness, and lifespan are all concerns with rechargeable battery technologies available for electric vehicles today.

Presently, hybrid vehicle makers tend towards incorporating nickel metal-hydride (NiMH) and lithium-ion (Li-ion) batteries into their products. Toyota's Prius, for instance, uses a sealed NiMH battery pack to supply power to the car's electric motor. When compared to the lithium-ion battery, the NiMH's power level is lower and the self-discharge rate is higher. With a shelf life of just three years, NiMH is not the ideal solution for electric vehicles.

Traditional lithium-ion batteries (Figure 3.6) offer high specific energy and low weight. However, due to their high cost, intolerance of temperature extremes, and safety – with safety being the single most significant hurdle in adopting Li-ion batteries today – these batteries are not the ideal solution either. In fact, Toyota recently decided to delay (by one to two years) the launch of new high-mileage hybrids with Li-ion batteries due to their concerns over safety.

Traditional Li-ion batteries offer a 3–5 year lifespan with 1000 cycles. Another issue is the size of the Li-ion battery necessary to guarantee the desired range in a vehicle for an acceptable battery lifespan. While we accept shorter than advertised run-times on our laptops and cell phones (as we

Figure 3.6
A Li-ion accumulator (Museum Autovision Altlussheim, Germany). Reproduced from wikipedia.org, with permission.

can simply plug them in while in use) we do not have this luxury with a moving vehicle. A battery for a hybrid vehicle would need to charge and discharge rapidly, so that a short charging time and acceptable acceleration can be achieved. An electric car battery has to be light, small, energy dense, and quick to recharge. But it also has to be relatively cheap, long lasting, and safe.

Financially supported by federal funds, researchers in the US in 2008 developed a lithium iron phosphate electrode material that achieves ultra-high discharge rates, comparable to those of supercapacitors, while maintaining the high energy density characteristic of lithium-ion batteries [4].

Lithium-ion batteries are presently popular in cell phones and laptops because they store a considerable amount of energy into a small, lightweight cell that can be recharged hundreds of times and holds its charge when idle. The power capability of a lithium battery depends heavily on the rate at which the ions and electrons can move through the active electrode material. Much of the work on improving the power rate for lithium-ion batteries has focused on improving electron transport in the bulk or at the surface of the material, or on reducing the path length over which the electron and the Li^+ ion have to move by using nano-sized materials. However, scaling these up for electric cars, for instance, requires a dramatic improvement in charge and discharge rates – otherwise a plug-in electric vehicle would need to be plugged in for many hours to fully recharge.

Gerbrand Ceder and colleagues at the MIT predicted that the material's lithium ions should actually be moving extremely quickly, into the material but *only* through tunnels accessed from the surface. Using nanochemistry, they thus created a new surface structure by creating a glassy lithium phosphate coating on the surface of nanoscale $LiFePO_4$ (Figure 3.7) that allows the lithium ions to move quickly around the outside of the material.

Glassy lithium phosphates are known to be good (and stable) Li^+ conductors. Now, when an ion traveling across this material reaches a tunnel it is instantly diverted into it, enabling the fast charge and discharge capability; in addition, further tests showed that unlike other battery materials the new material does not degrade as much when repeatedly charged and recharged.

The new material has a rate capability equivalent to full battery discharge in 10–20 s, namely 100 times faster than a regular rechargeable. Only 360 W is required to charge a 1 Wh cell phone battery in 10 s (at a 360 C charging rate). In contrast, the rate at which very large batteries such as those planned for plug-in hybrid electric vehicles can be charged is limited by the available power: 180 kW is needed to charge a 15 kWh battery (a typical size estimated for a plug-in hybrid electric vehicle) in 5 min (Figure 3.8). This extremely high rate capability of the electrode materials blurs the distinction between supercapacitors and batteries. By simply tweaking the composition of the battery material, nanochemistry enables fast charging and discharging rates – like a supercapacitor – while keeping the energy density comparable to that of a battery.

Figure 3.7
The LiFePO$_4$ nanoparticles are less than 50 nm in size. Reproduced from Reference [4], with permission.

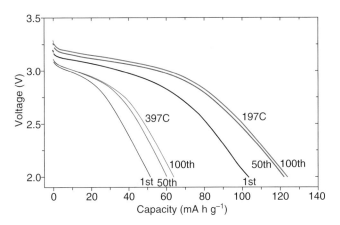

Figure 3.8
Discharge capability at very high rate for LiFe$_{0.9}$P$_{0.95}$O$_{4-\delta}$. Full charge–discharge cycles at constant 197 and 397 C current rates without holding the voltage. The 1st, 50th, and 100th discharges are shown for each rate. Reproduced from Reference [4], with permission.

The power density based on the measured volume of the electrode film, including carbon and binder, is around 65 kW l^{-1} in the 400 C test. Assuming that the cathode film takes up about 40% of the volume of a complete cell, this will give a power density of ~25 kW l^{-1} at the battery level, which is similar to or higher than the power density in a supercapacitor, yet with a specific energy and energy density one to two orders of magnitude higher.

The ability to charge and discharge batteries in a matter of seconds rather than hours may make possible new technological applications and induce lifestyle changes. Such changes may first take place in the use of small devices, where the total amount of energy stored is small. The technology has been licensed by two companies with the aim of manufacturing smaller, lighter batteries that are rapidly rechargeable and long lasting. In addition, since the material involved is not new, the work could make it into the marketplace within two to three years.

Meanwhile, the US company Altair Nanotechnologies has developed an innovative rechargeable battery, based on nanotechnology, that can meet advanced requirements in terms of short recharge time, long cycle life, complete operational safety, and tolerance for extreme temperatures (Figure 3.9). The technology is based on a nano-size lithium titanate oxide (nLTO) battery electrode material where nLTO substitutes for graphite, the standard negative electrode material employed in common Li-ion rechargeable batteries. These nano-titanate based batteries have one-third the weight and four times the power for the same sized NiMH battery.

Manufactured by Phoenix Motorcars, the electric SUV (Figure 3.10) is one of the first commercially available vehicles using Altairnano batteries. The vehicle is planned to be capable of road speeds up to 95 mph, a driving

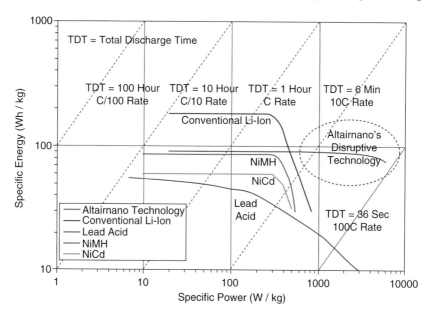

Figure 3.9
Altair Nanotechnologies is developing technology that operates in the high power region previously not served by other technologies. Reproduced from altairnano.com, with permission.

Figure 3.10
This electric SUV is a commercially available electric vehicle, from Phoenix Motorcars, that makes use of advanced lithium ion batteries. Reproduced from phoenixmotorcars.com, with permission.

range that is configurable up to 200 miles, a battery recharge time potential of less than 10 min, and an acceleration of 0 to 60 mph in less than 10 s.

Once again, the company uses nanochemistry – and, in particular, a technique called spray hydrolysis – to synthesize a novel nano-structured lithium titanate anode material with a process that enable precise control over the nanomaterial properties, including surface area, morphology, particle size, and purity.

Owing to replacement of the graphite electrode materials found in negative electrodes of traditional Li-ion batteries with nLTO electrodes, there is up to 100 times more surface area available to the ions, which facilitates access to the active sites required for battery operation, and reduces the distance from the surface to the sites, all of which helps accelerate recharging and discharging [5].

The mechanical stress and strain caused by ions entering and exiting electrodes reduces the life of a battery. During the charging of conventional lithium-ion batteries, lithium atoms (the reduced lithium ions) deposit inside the anode and are then released on discharge. Graphite possesses a two-dimensional crystal structure. When the lithium ions enter or leave the anode, the graphite planes of the anode shift and strain to a 10% greater separation of the planes to accommodate the lithium ion's size. On discharge, these processes are reversed. Over the life of the battery, this repeated shifting and straining fatigues the graphite planes, which causes the graphite structures to fracture. These fractures cause a loss in electrical contact between the particles, thereby reducing the battery's capability and capacity, and, ultimately, its life.

The use of nano-size lithium titanate oxide instead of graphite makes nLTO a "zero strain" material. Thus, the material essentially does not change shape upon the entry and exit of a lithium ion into and from a

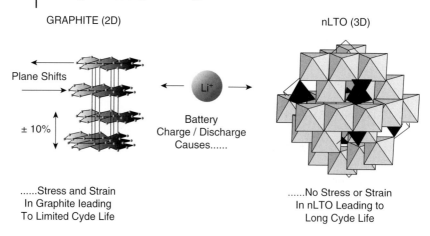

Figure 3.11 Relative mechanical processes in graphite and nLTO crystals during battery operation. Reproduced from altairnano.com, with permission.

particle. The nLTO material has a three-dimensional crystal structure. The structure (Figure 3.11) contains sites for lithium atom inclusion that are roughly the same size as the atom. Therefore, there is virtually no stress or strain involved with the charge and discharge process. Thus the battery can be charged and discharged significantly more often than conventional lithium-ion batteries because of the absence of particle fatigue, which plagues materials such as graphite. Conventional lithium-ion batteries can be typically charged about 1000 times before they are no longer useful (about 3 years), whereas cells using nLTO materials can achieve over 25 000 charge and discharge cycles (more than 20 years). Finally, these nanotech batteries can be safely operated at temperatures up to 240 °C – more than 100 °C above the temperature at which graphite-based batteries can explode – with zero explosion or safety concerns. Several Li-ion batteries blew up after overheating in laptops in 2006, proving that that the technology at that stage was simply not ready to be mass-produced for cars. Initial availability of these cells and advanced energy storage systems and batteries featuring the lithium-titanate cells has been anticipated by the end of 2009 following agreement between Altair Nanotechnologies and a manufacturer of lithium-ion battery cells for mobile devices (Amperex Technology).

3.3
Biological Fuel Cells

Fuel cells convert the energy stored in a fuel directly into usable electrical power. They are batteries that can be recharged by merely refilling a small fuel tank. One of the their main drawbacks is that they are limited to simple fuels such as hydrogen or methanol, both of which have serious safety concerns because hydrogen is explosive and methanol is flammable and highly toxic. Biofuel cells are fuel cells that use biocatalysts – and *no* expensive precious metals – to convert chemical energy into electrical energy; they are ideally suited to power portable devices using as fuel simple sugars that are readily obtained from biomass [6].

The US-based company Akermin has developed a technology called Stabilized Enzyme Biofuel Cell (SEBC) by which it immobilizes and stabilizes enzymes in a conductive polymer matrix and allows them to convert renewable organic fuels into electricity, a process that is inherently more efficient than conventional methods of portable power generation. The stability afforded by Akermin's enzyme immobilization polymers, measured in years compared to days for other biofuel cell technology, enables the first truly commercial viability for this type of power supply. The new biofuel cell technology uses a biocathode in which entrapped enzymes digest a the well-known syrupy substance glycerol – a safe sugar, containing three times more energy per liter than hydrogen, that is currently obtained in huge amounts as 10% by-product of biodiesel manufacturing (Figure 3.12).

Using such a high energy content fuel, these cells are more environmentally-friendly and last longer than any existing fuel cell, delivering lower cost per watt-hour over their lifetime.

Figure 3.12
Glycerol, a non-toxic substance abundantly obtained from biodiesel manufacturing, can be used to power efficiently laptops and mobile phones with a portable biofuel cell. Reproduced from commonvision.org, with permission.

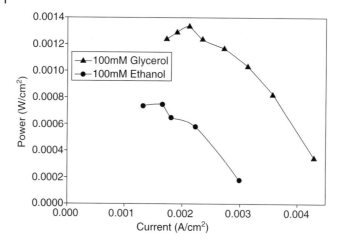

Figure 3.13
Comparison of the power curves for a single biofuel cell with two different analyte fuels (100 mM ethanol and 100 mM glycerol) at room temperature. Reproduced from Reference [7], with permission.

Glycerol furthermore can also be used in a biofuel cell at 98.9% concentration without damaging the cell, whereas methanol is limited to 40% concentration. As a result, the same amount of glycerol produces almost four times as much power as methanol and twice as much as ethanol (Figure 3.13), and is the main alternative being considered presently for portable electronics such as cell phones and laptops.

The technology makes use of membrane-immobilized enzymes [7]. Two oxidoreductase enzymes are immobilized at the surface of a carbon electrode in the pores of a Nafion ion exchange copolymer membrane chemically modified to expand its pores and make the environment more hydrophobic and enzyme-friendly – thereby ensuring stability of the enzymes for months or even years.

The resulting bioanodes have been incorporated into a glycerol–oxygen biofuel cell that enables multi-step oxidation of glycerol to mesoxalic acid; the overall process (Figure 3.14) utilizes 86% of the energy density of the glycerol and results in power densities of up to 1.21 mW cm^{-2} at room temperature. This is very different from metallic electrodes, which give glycerate as the only detectable oxidation product of glycerol, and it shows that the biofuel cell can allow deeper oxidation of the glycerol fuel, thereby increasing overall efficiency and energy density.

The technology was licensed to a start-up company called Akermin (www.akermin.com), which is working towards commercialization. Akermin has developed several prototypes (Figure 3.15) in low power ranges and has secured contracts with corporate and government entities to develop power supplies using the SEBC for commercial uses.

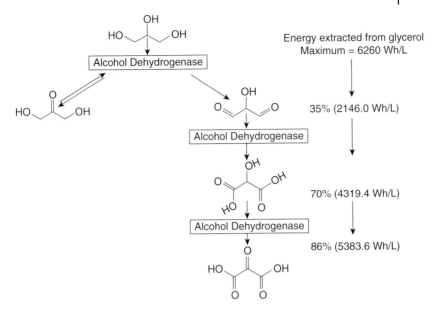

Figure 3.14
Oxidation sequence for glycerol at a PQQ-ADH/PQQ-AldDH-modified bioanode. Reproduced from Reference [7], with permission.

Figure 3.15
Akermin has developed several "world-first" prototypes in low power ranges and has secured contracts with corporate and government entities to develop power supplies using the SEBC for commercial and other uses. Reproduced from akermin.com, with permission.

The key to commercial development lies in improving the lifetime and performance of the entrapped enzymes over a range of temperatures. According to the inventor, Shelley Minteer at Missouri Saint Louis University, it will take about 2–4 years before glycerol clips could be used to power mobile phones. The company has, in particular, at present prototypes for powering wireless sensor networks that are being tested by several companies and is making extensive efforts to commercialize their air breathing biocathode for both biofuel cell applications and other electrochemical energy conversion devices (other types of batteries).

3.4
Fuel Cells for the People

In 2006 and again in 2009 central and western European countries, massive consumers of Russia's natural gas, were left for about four weeks without supply of gas as the Russian gas distribution company decided to end distribution, lamenting delayed payments from one customer country [8]. Most of this natural gas gets burned in advanced "turbogas" 55%-efficient thermoelectric power plants and the rest is burned at home by citizens either directly for cooking or in small heaters to produce sanitized water.

However, there is definitely a better way – much more efficient and less polluting – to use all this gas and extract the abundant chemical energy of methane and to convert it into electricity: namely to oxidize it in a controlled electrochemical process taking place in a fuel cell. This commercially long-awaited technology, originally developed to produce water and electricity onboard space craft in the late 1960s, has the potential to revolutionize both transportation and power generation, and is, finally, reaching the market.

Figure 3.16, for example, shows the 300 kW fuel cell stack that has supplied with both power and heat (hot water) to the university hospital in the city of Magdeburg in Germany since 2006 (www.mtu-friedrichshafen.com). In contrast to conventional cogeneration plants, a fuel cell power plant has a considerably higher efficiency and a much lower environmental impact. Emissions from the MTU system, for instance, are so low that, according to the German clean air code, the emissions can be classed as "exhaust air" instead of "exhaust gas," with exhaust air consisting mainly of hot air, carbon dioxide, and water vapor.

Other MTU fuel cell power plants are operating in Germany at a Michelin tire plant in Karlsruhe, a telecommunications center in Munich, and a hospital in Bad Neustadt/Saale, as well as at a shipbuilding facility in Cartagena, Spain. In Germany the consortium MTU, which includes FuelCell Energy, the electric utility RWE, and Daimler, was established to market stationary carbonate fuel cells, starting with the many companies and public bodies that are customers of RWE.

Figure 3.16
The long life and reliability of the Hot-Module was proven over nearly four years of operation at the University Clinic in Magdeburg. The fuel cell achieved a remarkable 98% availability over 30 000 hours in operation and there were no measurable signs of aging at the scheduled end of the project. Reproduced from Hfpeurope.org, with permission.

New generation fuel cells, however, will only run on hydrogen produced by solar water splitting, thereby eliminating carbon emissions (Figure 3.17). In particular, these H_2-based fuel cells onboard a vehicle will power the electromagnetic engine propelling the vehicle. For example, Class 212 and 214 submarines (Figure 3.18) of the German, Italian, and Greek Navies are equipped with powerful fuel cells manufactured in Italy by Ansaldo Energia for long-duration, silent travel. Electric propulsion here is strategic as the electromagnetic engine of the submarine has the lowest noise emission; so low that it cannot be detected by enemy sonar.

The first in the world to be equipped with an ultramodern hydrogen fuel cell engine, this submarine can operate submerged for weeks independently of external air. Electrochemical oxidation of hydrogen indeed produces pure water that is used for the crew's needs:

$$H_2 + \tfrac{1}{2} O_2 \rightarrow H_2O$$

Clearly, the sustainability crisis requires fuel cell technology to exit the laboratory and military or space applications, where cost often is not a matter of concern, to become a cost competitive energy generating technology for ordinary people. The new fuel cell commercialized by the Canadian company Ballard (Figure 3.19) for instance was designed for 1–5 kW home applications. It can thus be used to provide a regular house with its electricity needs.

Figure 3.17
The hydrogen-fueled 5 kW Penta fuel-cell powering the HydroLAb building in Italy makes use of renewable hydrogen only. Reproduced from lafabbricadelsole.it, with permission.

Figure 3.18
This 1450t Class 212 A submarine of the Greek Navy is powered by hydrogen fuel cells. The fuel cells are used for long-duration operations at moderate speeds, while a diesel–electric engine with batteries is used for high-speed maneuvers. Reproduced from evworld.com, with permission.

Figure 3.19
The Ebara Ballard 1 kW cogeneration system includes a Ballard fuel cell stack, Ebara Corporation pumps, a reformer, and a hot water storage tank. This generator has achieved a total system efficiency (heat and electricity) of 92%. Reproduced from ballard.com, with permission.

One of the leading fuel cell technologies developed, in particular for transportation applications, is the proton exchange membrane (PEM) fuel cell. These fuel cells are powered by the electrochemical oxidation reaction of hydrogen and by the reduction of the oxygen contained in air. These fuel cells run at relatively low temperature (<100 °C) and therefore need catalysts to generate useful currents at high potential, especially at the electrode where oxygen is reduced (the cathode of the fuel cell). The key technology aspect thus lies in the development of high-performing and resistant solid catalysts to assist the (sluggish) oxidation reaction.

Presently, platinum-based electrocatalysts are the most widely used in PEM fuel cell, such as, for example, in the thirty-six fuel cell buses (Figure 3.20) operated on roads in Europe, Australia, China, and the United States using PEM cells manufactured by the Canadian company Ballard Power Systems. Platinum of course is very expensive due to its limited supply and its price is also highly volatile, thereby creating one of the major barriers preventing commercialization of PEMs.

The estimated cost for a gasoline engine is about $30 per kilowatt (kW). According to the US Department of Energy, the system cost for automotive fuel cells has gone from $275 per kW in 2002 to $95 per kW in 2008 and is projected to be $60 per kW in 2009. The target is $30 per kW by 2015. The same department has established some performance targets for

Figure 3.20
Thirty-six Ballard-powered fuel cell buses have operated on roads in Europe, Australia, China, and the United States. Vehicles like this bus in London are providing Ballard with valuable information about how fuel cell powered buses perform in different geographies and climates. Reproduced from ballard.com, with permission.

platinum use in PEM fuel cells for automotive application: 0.3 and 0.2 g of platinum per kW of PEM fuel cell stack for 2010 and 2015, respectively; this will require nanomanipulation of the platinum catalyst to improve the specific activity.

Recently, Jean-Pol Dodelet and his team at Canada's Université du Québec has shown that this is entirely possible by using platinum nanowires as electrocatalyst instead of the usual platinum nanoparticles (Figure 3.21).

A simple room temperature anisotropic growth method enables the aqueous phase synthesis of single-crystal nanowires of platinum on the nanospheres of a carbon black (a commonly used catalyst support in fuel cells). The resulting nanostructures–with the high-surface-area carbon black as the core and the electrocatalytically active platinum nanowires growing radially from the surface of the carbon particles–show greatly enhanced catalytic activity for the oxidation reaction compared with a state-of-the-art platinum/carbon catalyst made of platinum nanoparticles [9].

In particular, the platinum nanowire catalyst shows a 50% higher mass activity than the commercial cathode despite a 50% lower platinum area for the platinum nanowire catalyst. In other words, taking into account both effects, the specific activity of the platinum nanowire catalyst is three-fold better than that of the commercial cathode. These Pt nanowires will soon replace spherical platinum nanoparticles used as catalysts in

Figure 3.21
(a) Scanning electron microscopy (SEM) image of platinum nanostructures synthesized via a simple wet chemical method, at room temperature, using neither template nor surfactant. The nanostructures consist of numerous single-crystal Pt nanowires with diameters of ca. 4 nm and lengths that may reach 100 nm. (b) Detailed view of (a). Images courtesy of nanowerk.com, Dr. Dodelet/INRS.

first-generation commercial PEM fuel cells. Once again, it is seen that nanochemistry enables progress of practical interest.

References

1 Rüdiger, J. (2002) *Die Quandts*, Campus Verlag, Frankfurt.
2 IPCC, IPCC Fourth Assessment Report (2008) *Climate Change 2007: Synthesis Report* (eds R.K. Pachauri, and A. Reisinger), IPPC, Geneva. Available at www.ipcc.ch/publications_and_data/ar4/syr/en/contents.html (Last line accessed October 30, 2009).
3 Beigbeder, F. (2003) *Was 9.99, Now 6.99*, Picador, New York.
4 Kang, B. and Ceder, G. (2009) Battery materials for ultrafast charging and discharging. *Nature*, **458**, 190.
5 House, V. and Ross, F. (2007) How to build a battery that lasts longer than a car. *Power Management DesignLine* (Aug 27). http://www.powermanagementdesignline.com/showArticle.jhtml;jsessionid=PUCU10BLGMMHWQSNDLRSKHSCJUNN2JVN?articleID=201802331.
6 These cells will be used between freezing temperatures and 90 °C, and so they will not be used in high temperature applications like automobiles and household generators: Cooney, M.J., Svoboda, V., Lau, C., Martin, G., Minteer, S.D. (2008) Enzyme catalysed biofuel cells. *Energy Environ. Sci.*, **1**, 320.
7 Arechederra, R.L., Treu, B.L., and Minteer, S.D. (2007) Development of glycerol/O_2 biofuel cell. *J. Power Sources*, **173**, 156.
8 Philip, P.P. (2009) Russian fuel cuts felt across Europe. The Washington Post (Jan 7).
9 Sun, S., Jaouen, F., and Dodelet, J.-P. (2008) Controlled growth of Pt nanowires on carbon nanospheres and their enhanced performance as electrocatalysts in PEM fuel cells. *Adv. Mater.*, **20**, 3900.

4
Catalysis: Greening the Pharma Industry

4.1
Pharma: An Industry to Be Cleaned Up

The pharmaceutical industry is among the world's largest industries in terms of turnover, with revenues of more than $600 billion in 2008. The industry certainly plays a central role in maintaining our health [1], but it does so generally by producing between 25 and 100 kg or more of waste for every kilogram of active pharmaceutical ingredient (API) manufactured. For comparison, the petrochemicals sector produces 0.1 kg of waste for every kilogram of product produced. Of course, commercial volumes of drugs are much lower, with an annual production of between 1000 to 1 million kg per compound, compared with basic chemicals that are produced in billions of kilograms per year. Nevertheless, even at a nominal disposal cost of $1 per kg, the potential savings just in waste avoidance is significant, being in the range of 500 million to 2 billion kg per year [2].

API molecular structures are generally complex, the syntheses are lengthy, and patient safety demands very high purity. In addition, unlike their chemical-sector neighbors, most APIs are made in a batch mode rather than by continuous processing, using obsolete chemical conversions in solution that consume large amounts of solvents and heavy metals dissolved therein that function as key reactants. By applying the "green chemistry" principles [3] (Figure 4.1) to redesigning existing manufacturing processes and to designing processes for new drugs, waste reductions of more than tenfold have been achieved. These reductions provide a *double* economic benefit to companies because more of the raw materials they purchase end up in the products and less waste needs to be disposed of.

Here, again, major advances have been made in the last few years that are generally in the field of nanoscale chemistry science and technology, affording several key chemical technologies whose potential, in terms of benefits to industry, the environment and society, is finally being realized.

Nano-Age: How Nanotechnology Changes our Future. Mario Pagliaro
© 2010 WILEY-VCH Verlag GmbH & Co. KGaA, Weinheim
ISBN: 978-3-527-32676-1

Figure 4.1
The Green Chemistry "evangelist," American chemist Paul Anastas, published the manifesto of the new, clean chemistry in 2000. Image courtesy of Professor Paul Anastas.

How can we achieve the same end-product using a different set of inputs and reactions such that we eliminate waste from the process? By developing new, effective catalysts. Unlike other reagents that participate in a chemical reaction, a catalyst is a substance that is not consumed by the reaction itself. Its role, instead, is to enhance the rate of a desired chemical reaction. Eventually, by reducing waste and the costs associated with disposing of or recycling that waste, this approach to synthetic chemistry based on catalysis ends up saving companies money. In particular, wasteful "stoichiometric" processes are being replaced with clean catalytic conversions using solid catalysts. This allows elimination of the separation step of the catalyst from the reaction mixture, recovery of the costly catalyst, and fast consecutive conversions with easy separation from the products in the reaction mixture.

In a regulated industry where products and processes must be validated prior to receiving marketing approval it is much more difficult to change the process post-launch. Thus, by the time of product launch, the pharma industry aims to have already applied green chemistry practices wherever possible to minimize waste and environmental impact and to "reap the benefits of that process from day one of the launch" [4]. The German manufacturer Merck, for example, has developed a new process for producing sitagliptin—the active ingredient in Januvia, a drug for type 2 diabetes—that reduced the waste per kilogram of sitagliptin manufactured by 50%, and completely eliminated aqueous waste streams. The concept is general: new chemical industrial processes may now be conducted in a one-step, solvent-free manner, minimizing consumption of energy, pro-

Figure 4.2
Oxidation of hydrocarbons into highly useful oxygenate derivatives now takes place smoothly with oxygen using nanoporous catalysts.

duction of waste, and avoiding the use of corrosive, explosive, volatile, and non-biodegradable materials. All these requirements may eventually be achieved by designing the appropriate solid catalyst made of nanoporous solids that meet stringent demands in terms of sustainability, feasibility, and economic viability.

As another example, so-called C–H activation is of central importance throughout the fine chemicals industry and new heterogeneous catalysts have been developed successfully. Many highly selective oxyfunctionalized hydrocarbon products may now be produced using O_2 or air (or a solid source of active H_2O_2) as oxidants under environmentally benign conditions, often requiring no solvent [5]. A typical example is the conversion of styrene into styrene oxide (Figure 4.2).

No protection groups are required for these industrially significant conversions, but advantage is taken of single-site heterogeneous catalysts to achieve the required selectivity. The trick, once again, lies in our new ability enabled by nanochemistry advances to design nanostructured solids in which well-characterized active centers are anchored to the framework of the inner walls. The very general concept is analogous to the one-century old Fisher's "key-in-the-lock" principle that explains the selectivity of enzymatic reactions, and holds large validity also in heterogeneous catalysis for fine chemicals synthesis within the nanoporosity of solid materials (Figure 4.3) [6].

Clearly, at the concave surface, access of the substrate to the catalyst is favored only when access takes place along the axis of the pore; whereas no such restrictions exist in the case of the same catalyst anchored at the convex surface. This explains why in the first case the selectivity is considerably higher and has opened the way to bringing chemistry to a confined space.

Eventually, chemical syntheses will be carried out with the same kind of work flow as happens with the modern construction of a car, by performing catalytic reactions safely and with greater control, one after another, in a continuous microreactor (Figure 4.4), namely in a small tube designed to remove or introduce heat from or into a reaction precisely and with great speed [7].

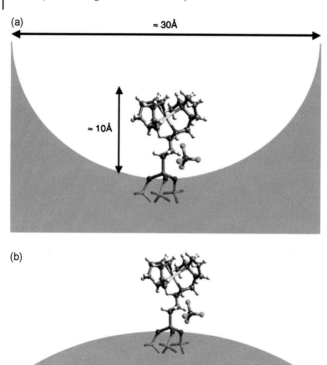

Figure 4.3
This graphical model of a catalyst anchored to a silica surface shows the constraints when the surface is concave (a) or convex (b). Reproduced from Reference [6], with permission.

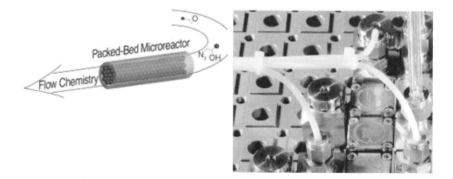

Figure 4.4
New chemical syntheses in industry will be carried out in microreactors equipped with catalytic nanomaterials performing catalytic reactions one after another until the desired product is obtained. Photograph courtesy of EPF Lausanne.

Microreactor technology is already an efficient tool for kilogram-scale syntheses in continuous mode and is particularly effective for hazardous reactions that do not allow scale-up in conventional reactors. Applications to several classes of reactions are described herein, including highly exothermic reactions, high-temperature reactions, reactions with unstable intermediates, and reactions involving hazardous reagents. Again, to find widespread use in industry, suitable nanocatalysts will have to be developed for such microreactors that couple together multiple catalytic transformations in series.

4.2
Sol-Gel Catalysts: Philosopher's Stones

Catalysis over efficient heterogenized catalysts enabling one-pot, multistep synthesis offers in principle a solution to most of the problems encountered in pharmaceutical syntheses [8]. The large amounts of solvents and purification media currently employed by industry would be eliminated along with the lengthy processes that originate them. Yet, until recently heterogeneous catalysts in the fine chemicals industry were remarkable for their absence despite the fact that until the early 2000s many homogeneous catalytic systems could not be commercialized because of difficulties associated with separating the products from the catalyst. This obsolescence was mostly rooted in the nature of the fine chemicals industry, which is a product (and not a process) oriented industry, that is, it focuses on the development of new products to maximize revenues in the time span in which exclusive rights are granted by patenting innovation [9].

As a result, traditional heterogenization technologies of homogeneous catalysts were characterized by a poor level of performance in terms of activity, selectivity, and stability. Current economic hypercompetition, however, and ever stricter environmental regulations are bringing about a radical change in industry. Companies went back to academia to work together on the development of various solid catalysts for high-throughput organic synthesis. Catalysis by sol-gel "doped" materials – porous metal oxides confining an active species – has emerged in the last 10 years as a prominent tool to synthesize a vast number of useful molecules both in the laboratory and in industrial plants [10]. The term "sol-gel" generally refers to a low-temperature method using chemical precursors that can produce glasses with higher purity and better homogeneity than high-temperature conventional processes, in various shapes, including powders, fibers, coatings and thin films, monoliths, and porous membranes.

The underlying basic concept to all applications is unique and was first conceived in 1984 by the Hebrew University of Jerusalem's scientist David Avnir (Figure 4.5). One or more host molecules are entrapped by a mild process in a liquid phase within the cages of silica glass, where they are

Figure 4.5
Israeli scientist David Avnir, a chemistry professor at The Hebrew University of Jerusalem, has established the field of sol-gel organic hybrid materials. Photograph courtesy of Mario Pagliaro.

accessible to diffusible reactants through the inner pore network and give rise to chemical interactions and reactions [11]. The sol-gel entrapment largely enhances both the physical and chemical stability of the dopant, whereas the overall materials act as a chemical sponge: chromatographic materials that adsorb and concentrate the reagents at the cages surface where reactions take place.

Following the 1984 discovery, it was rapidly established that *any* organic molecule, including enzymes, could be entrapped and dispersed within the *inner* porosity of such glasses with full retention of the chemical activity and marked stabilization of the entrapped dopant molecules. The domains of organic chemistry and ceramic materials were merging and the new era of inorganic–organic hybrid materials had started [12]. Despite its simplicity, the concept has tremendous implications and the field has rapidly evolved into the so-called "sol-gel science and technology:" an advanced domain of chemical research capable of affording sophisticated catalytic materials that offer unique versatility and many new useful properties.

Sol-gel catalysts are heterogeneous materials (Figure 4.6) in which a mobile and a stationary component penetrate each other at the molecular level, with the catalytic species being well-defined, highly *mobile* and *homogeneously* distributed across a highly porous chemically and thermally inert network [13]. They thus combine the advantages of homogeneous (high selective activity and reproducibility) and heterogeneous (stabilization and easy separation and recovery of the catalyst) catalysis. In general, therefore,

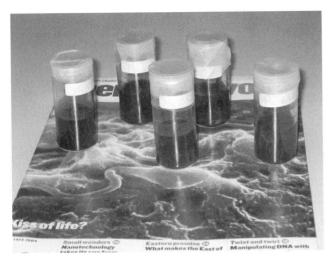

Figure 4.6
The alcogels in this picture are ORMOSILs doped with a Ru molecular species called TPAP. Upon a mild heat treatment these materials become more active than TPAP in solution. Photograph courtesy of the author.

a sol-gel entrapped catalyst shows higher selectivity than either a homogeneous catalyst in solution or a catalyst surface-bonded to a non-porous material.

In most applications, silica is the matrix of choice. This is because silica has several advantageous properties that make it an optimal commercial support for solid-phase syntheses, including stability towards harsh conditions, low swelling, and consistent binding sites for the catalyst (Box 4.1). The Canadian based company SiliCycle Inc., for instance, commercializes a vast variety of functionalized silica gels as advanced reagents for the pharmaceutical and fine chemicals industries.

Even better, instead of silica as such, sol-gel organically modified silicates (ORMOSILs) have shown enormous potential as catalysts for selective conversions in advanced organic chemistry. In these nanohybrid glasses an intimate mixing of molecular alkoxide precursors in organic solvents is realized that allows organic and inorganic components to be associated at the molecular level to form real organic–inorganic nanocomposites.

For example, the versatile aerobic oxidation catalyst TPAP (tetra-n-propylammonium perruthenate) – which shows modest activity in promoting alcohols oxidation when entrapped in a pure SiO_2 matrix – entrapped in nanoporous partially hydrophobized silica (Figure 4.7) is far *more* active than the catalyst dissolved in toluene (Figure 4.8) [14]. Given the industrial relevance of catalysts using oxygen as a unique "aerobic" oxidant, the catalyst was commercialized by SiliCycle under the trademark SiliaCat® PerRu.

4 Catalysis: Greening the Pharma Industry

> **Box 4.1 Advantages of using silica as catalyst support.**
>
> **Fast kinetics**
>
> Silica is surface functionalized and reacts much faster than conventional polymer bound reagents where the reaction is slowed by the rate of diffusion through the polymer and can be slowed further by the polymer's ability to swell.
>
> **Versatility**
>
> Silica works under a wide range of conditions: in all solvents, organic and aqueous. It has a high thermal stability and can be used in microwave applications.
>
> **Ease of use**
>
> Unlike polymers, silica is easy to weigh and handle with no static issues and is easily amendable to automation. It is mechanically stable, works in any format, and is easily scaled. It requires little or no washing because it does not swell in any solvent.

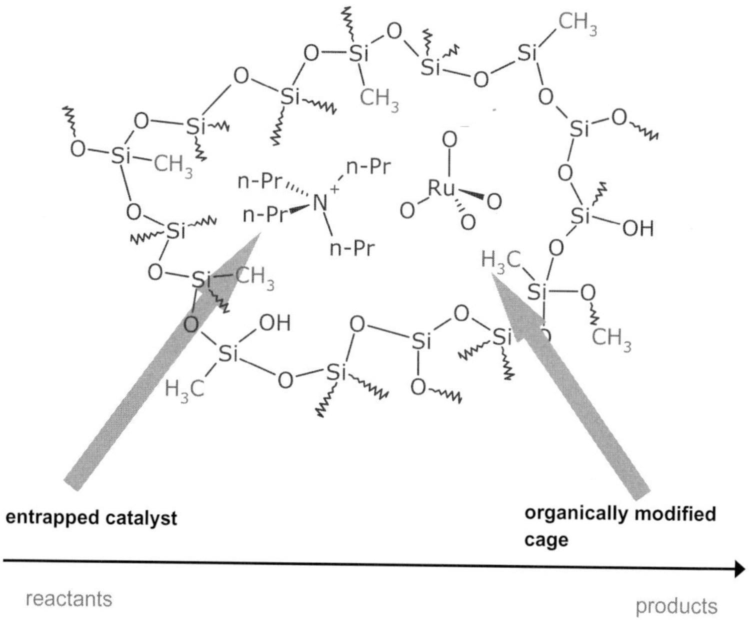

Figure 4.7
Sol-gel caging of a dopant (TPAP, in this case) is versatile. In addition, a methylated ORMOSIL such as the one represented here is an ideal aerobic catalyst. Image courtesy of Mario Pagliaro.

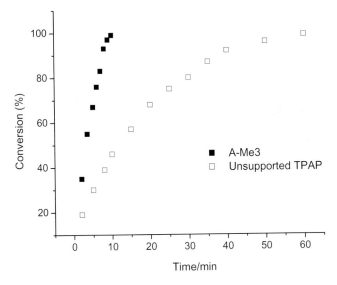

Figure 4.8
Oxidation kinetics in the aerobic conversion of benzyl alcohol into benzaldehyde in toluene mediated by 10 mol.% TPAP encapsulated in the sol-gel hydrophobic matrix A-Me3 (■) and unsupported (□). Reproduced from Reference [14], with permission.

The first sol-gel commercial catalyst reached the market in 2005 thanks to the Swiss company Fluka, which commercialized hybrid glasses doped with the industrially well-known lipase class of enzymes (see below). A more recent example is Silia*Cat* TEMPO (TEMPO = 2,2,6,6-tetramethyl-piperidine-1-oxyl), a metal-free "off-the-shelf" solid catalyst made of hybrid silica suitable for the conversion of alcohols into delicate fragrances or carboxylic acids. Fragrances (Figure 4.9) are generally made of aldehydic molecules, and especially unsaturated molecules. The catalyst Silia*Cat* TEMPO can smoothly afford large yields of aldehydes without the serious drawbacks in terms of selectivity, metallic wastes, safety, and harsh conditions posed by older processes used for this fundamental transformation (Figure 4.10).

As stated in the 1920s by Ernest Beaux, the perfumer who created Chanel n°5, in perfumery "the future lies primarily in the hands of the chemists;" indeed, it is the chemist who finds new fragrances (aroma chemicals) for both perfumery and personal care products by synthesizing and thus creating new, original notes.

Today, nano-encapsulation of the highly reactive radical species TEMPO has enabled the clean, rapid synthesis of various fragrances under extremely mild conditions (4°C) in a biphasic water/solvent mixture, with complete recovery and recycle of the Silia*Cat* TEMPO catalyst at the end of the

SiliaCat TEMPO

Some aldehydes used in fragrances

$H_3C-(CH_2)_8-CHO$ $H_3C-(CH_2)_8-\underset{CH_3}{CH}-CHO$ $Ph-CH=\underset{CHO}{C}-(CH_2)_5-CH_3$

1-decanal — orange, lime

2-methylundecanal — citrus, fruity

2-hexylbenzene-2-propenal — strawberry, peach

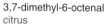

3,7-dimethyl-6-octenal
citrus

trans-2-hexanal
strawberry

isohexenylcyclohexenyl carboxaldehyde

orange

Figure 4.9
SiliaCatTEMPO: an off-the-shelf oxidation catalyst yielding valuable aldehydes, ketones, or carboxylates, depending on the reaction conditions (adapted from figures of ARKEMA*).

Oxidation of alcohols into aldehydes

$R-CH_2-OH \longrightarrow R-C\underset{H}{\overset{O}{\diagup\!\!\!\diagdown}}$

Stoichiometric reactions
Collins or Sarret reagent : CrO3/pyridine
Corey reagent : CrO3/pyridine/HCl (PCC)
Swern : oxalyl chloride/DMSO
Corey-Kim : DiMethylSulfide/N-Chloro-Succinimide
Dess-Martin periodinane (DMP)
Others : SO3/pyridine, KMnO4, MnO2, RuO4...

Catalytic dehydrogenation
Copper chromite (gas phase, high temperatures)
Others: Ni Raney, Pd(OAc)2...

Figure 4.10
Traditional oxidation processes employed in industry to obtain fragrances generally made use of hazardous and toxic reactants (adapted from figures of ARKEMA*).

* ARKEMA (document can be accessed at http://www.arkema.com/pdf/EN/products/research_and_development/Oxynitrox/oxynitrox_v3_presentation.pdf).

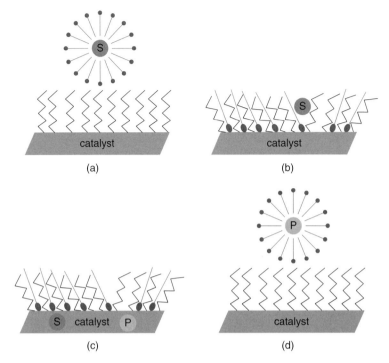

Figure 4.11
Illustration of the transport, reaction, and adsorption/desorption steps of the EST process. The emulsion which contains the substrate (a) spills its content into the catalyst material (b), the catalytic process takes place (c), and then the adsorbed surfactant carries the product back into solution (d). Reproduced from Reference [16], with permission.

reaction. The sol-gel encapsulation dictates accessibility of the alcohol molecules to the entrapped catalyst, limiting oxidative cleavage, and at the same time preventing catalyst deactivation (which is due to intermolecular quenching of the radical species bound at the silica surface) [15].

Similarly, the employment of catalytic doped ORMOSILs in water and soap opens the way to using water as the reaction medium for fine chemistry organic reactions by simply emulsifying the organic reactants, which typically are all hydrophobic, in an elegant three-phase emulsion–solution–transfer (EST) process (Figure 4.11) [16]. The surfactant molecules, which carry the hydrophobic substrate, adsorb/desorb reversibly on the surface of the sol-gel matrix, breaking the micellar structure and spilling their substrate load into the sol-gel cages where the catalytic reaction takes place to form the desired product, which is then extracted by the desorbing surfactant carrying the emulsified product back into solution.

4.3
Biogels: Marriage of Glass and Life [17]

Showing its unmatched potential in biomaterials synthesis, the sol-gel process can be efficiently applied for the encapsulation in silica glasses of practically *any* biological species, including enzymes, whole cells, antibodies, and even bacteria. This is an almost incredible result considering the amount of toxic alcohols that is released in conventional sol-gel processes based on the hydrolysis of alkoxides.

In general, the mild preparation conditions of ORMOSIL are compatible with the effective entrapment of transition metal catalyst, enzymes, or even living cells in silica-based materials (with minor or no loss of biological activity), that is, the *merger* of chemistry, biology, and materials science. The methodology has improved so much that today it is possible to encapsulate the most delicate of these giant molecules without reducing the original enzyme activity. In this manner, enzymes and cells are easily immobilized in biosilica gels, which display activities approaching those of the free biologicals, together with the high stabilities and robustness that characterize sol-gel bioceramics.

Shortly after the 1994 patent [18] claiming their invention, German chemist Manfred Reetz (Figure 4.12) discovered in 1995 that the enzyme

Figure 4.12
German chemist Manfred Reetz discovered in 1995 that enzyme lipase entrapped in ORMOSIL nanomaterials is up to eight times more active than the free enzymes. Nanochemistry alters and improves traditional chemistry in solution. Photograph courtesy of the Royal Society of Chemistry.

lipase entrapped in ORMOSIL nanomaterials was eight-times more active in esterification reactions than the enzyme powders used in organic solvent. Moreover, the materials chemical stability was extremely high, affording repeated usage of the same material in consecutive reaction cycles. Lipases are the most widely used enzymes in industry. Hence, the chemicals manufacturer Fluka rapidly commercialized a vast set of catalytic sol-gel lipase immobilizates. The materials have now reached a second-generation level of performance and are the blockbuster of the company, generating yearly revenues of several tens of millions of euros [19].

These spongy materials can host a large amount of catalytic molecules, leading to desirable high S/C (substrate/catalyst ratio). This, in practice, means that a very small amount of material can convert a large amount of substrate (e.g., only 250 mg of sol-gel lipase immobilizate is needed to convert 10 g of substrate in a typical preparative-scale reaction), making the biological process based on sol-gel catalyst largely competitive with the commercial BASF process at a scale of 1000 tons per year.

Several drugs and other industrially relevant molecules, including valued paclitaxel (Taxol), are currently being produced over cells entrapped in a thin siliceous layer; the cells are entrapped, thanks to the Biosil method, by spraying an aqueous silicate solution of a substance known as TEOS (tetraethyl-orthosilicate) over a suspension of living cells (Figure 4.13) [20].

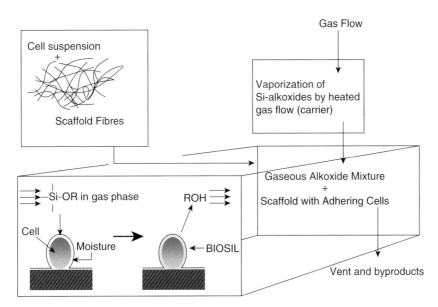

Figure 4.13
Biosil process: Si–OR → alkoxide precursors, ROH → reaction by-products. Reproduced from Reference [20], with permission.

Developed by Italian chemist Giovanni Carturan at the University of Trento in 1989 this approach – the use of functional cells or cell aggregates as a highly specialized laboratory for producing the desired substances – has enormous practical consequences.

Cells indeed are a complete and natural system encompassing *per se* the entire enzymatic chain and yielding specific molecules from elemental substrates. Furthermore, the process is general and versatile: gas-phase silicon alkoxides such as TEOS react with the wet surface of cells, affording a mechanically stable and homogeneous layer of amorphous SiO_2; the layer does not suppress cell viability or functionality while, allowing porosity control, it provides important immunological protection (by tailored exclusion of access to macromolecules above a certain threshold pore size).

Exploiting this technology, the Italian-based company IRB (Istituto di ricerche biotecnologiche; www.irbtech.com) has licensed the technology and started production of various biologically active molecules over sol-gel entrapped cells of Taxus brevifolia. In particular, IRB has optimized an industrial culture method called HTN (High Tech Nature) to produce active ingredients from plant cell cultures under strictly controlled conditions; the ingredients are obtained as secondary metabolites of the culture. By exceeding the limit of the plant biological cycle and maintaining its natural condition of growth and development, the process (Figure 4.14) considerably reduces production cycle times.

For example, verbascoside and isoverbascoside are substances with high biological activities obtained from Syringa vulgaris L., also known as the Common Lilac, a deciduous shrub from the Oleaceae family that is now found throughout Europe, where it is widely used as an ornamental garden plant. Verbascoside has a strong antioxidant activity – higher than that of common natural antioxidant benchmarks; it provides protection against skin functional disorders (acne, dermatitis) and hair oxidative stress, and prevents skin aging. Commercialized as a nutritional additive with the trademark Verbasyr (Figure 4.15) both the active principles are obtained by cultivating cells of the plant entrapped in the Biosil protection layer invented by Carturan.

The possibilities are so numerous that, in practice, today's pharmaceutical companies can now rely on sol-gel entrapped catalysts for both rapidly generating new drug candidates and screening these candidates for toxicity. In 2005 the group of Jonathan Dordick at New York's Rensselaer Polytechnic developed a device called a MetaChip (metabolizing enzyme toxicology assay chip) that combines high-throughput P450 catalysis with cell-based screening on a microscale platform so that many drug candidates can be tested simultaneously at early phases of drug development [21].

The technology provides a high-throughput microscale alternative to currently used *in vitro* methods for human metabolism and toxicology screening based on liver slices, cultured human hepatocytes, or isolated P450 itself. P450 enzymes are the liver's major detoxification enzymes:

Figure 4.14
Using sol-gel entrapped cells in a technology called High Tech Nature, the Italian biotech company IRB can manufacture high yields of highly pure organic molecules obtained as secondary metabolites of plant cells. Reproduced from irbtech.com, with permission.

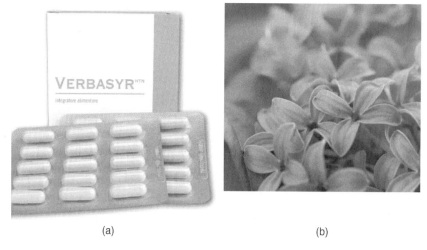

(a) (b)

Figure 4.15

(a) A 30-capsule package of Verbasyr administered as nutritional supplement; the actives contained in this natural anti-inflammatory are obtained over sol-gel entrapped cells of the Common Lilac (*Syringa vulgaris*) (b). Reproduced from irbtech.com, with permission.

iron-containing proteins responsible for the initial clearance of drugs from the body and the activation of prodrugs, which work by oxidizing chemicals to make them more water-soluble so that potentially harmful substances can be eliminated more easily from the body.

The chip is based on ORMOSIL-entrapped cytochrome P450 enzymes. The system allows us to create liver metabolites of drug candidates and to test them rapidly for toxicity against specific types of cell, identifying those activated by the liver and weeding out those made toxic earlier in the drug discovery process.

In practice, the device is made of sol-gel spots (varying in volume from 5 to 100 nl) of a precursor solution of recombinant P450 enzymes in organosilica arrayed over a glass slide (Figure 4.16). The second component is a human cell monolayer housed in a chamber slide, which is used for cytotoxicity screening of the P450-generated metabolite. A company, Solidus Biosciences (www.solidusbiosciences.com), was established in 2002 to enable cosmetics, chemical, and pharmaceutical companies to adopt this highly predictive sol-gel nanotechnology for *in vitro* toxicity screening to replace or minimize the need for animal testing. The company currently performs toxicology assays for customers using MetaChip platforms specifically fabricated to accommodate unique customer requirements, and is planning to sell MetaChips directly to customers in the near future.

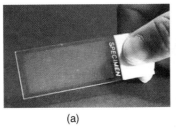
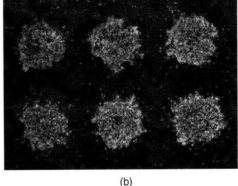

Figure 4.16
The MetaChip (a) can perform rapid toxicity testing of potential drug candidates. (b) A monolayer of stained human breast cancer cells after stamping onto the MetaChip. The colored spots correspond to regions of dead cells resulting from contact with a toxic product of P450 metabolism generated on the MetaChip. Reproduced from Reference [21], with permission.

4.4
Nanocatalysts: Abating Polluting Emissions and Product Contamination

Zero emissions is the true quality standard for environmental performance. In other words, our ultimate aim must be to replace traditional technologies currently employed to manufacture goods, or to produce kinetic or electric energy, with clean technologies affording innocuous by-products such as water. While this is being accomplished, however, we must continue to reduce damage to the environment by abating pollutants that continue to be produced, such as in the case of emissions from automobile engines. Here, nanochemistry helps again, affording second-generation nanocatalysts for automobile catalytic converters such as the single-nanocatalyst technology developed by Mazda (Figure 4.17).

The single-nanocatalyst in the under-floor catalytic converter in the Mazda3 requires only $0.15\,g\,l^{-1}$ of expensive precious metals such as platinum, palladium, and rhodium needed for three-way catalysts to effectively clean exhaust emissions from gasoline engines, which is approximately 70% less than the $0.55\,g\,l^{-1}$ required in the previous model. Even with the substantial reduction in precious metal usage, the 2010 Mazda3 meets the latest emissions regulations.

Automotive catalysts consist of a base material coated with precious metal particles. These metals promote chemical reactions that purify exhaust gases. In conventional catalysts, exposure to hot exhaust-gases causes the precious metal particles to agglomerate into larger clumps, which reduces their effective surface area and catalytic activity. To counteract this,

Figure 4.17
The model Mazda3 commercialized in 2009 by the Japanese car manufacturer is the world's first to use the single-nanocatalyst technology in its catalytic converter. Reproduced from mazda.com, with permission.

an increased amount of the precious metals is required to maintain an efficient purification performance.

Mazda developed the single-nanocatalyst to increase the effective surface area of the precious metals used.[1] By developing a method of controlling precious metal particles that are less than 5 nm in diameter as well as a proprietary catalyst material structure, Mazda created the first catalyst that features single-nanosized precious metal particles embedded in fixed positions into the base material (Figure 4.18). The new catalyst has two main features:

1) it inhibits the thermal deterioration caused by the agglomeration of precious metal particles;
2) it offers a significant improvement in oxygen absorption and release rates for enhanced emissions cleaning.

With these features, the amount of precious metals needed to ensure the same level of effectiveness is reduced by 70–90% compared to previous products. At the same time, the performance of the catalytic converter is almost unaffected by harsh driving styles. Mazda plans to replace its entire powertrain lineup and begin introducing the new product towards the start of the decade beginning in 2010.

The concept is general and has been extended by BASF to its NanoSelect platform technology for hydrogenation reactions (the partial reduction of functional groups) used in fine chemical and pharmaceutical applications.

1) http://www.mazda.com/mazdaspirit/env/engine/catalyst_technology2.html.

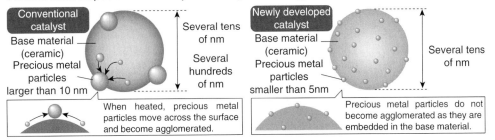

Figure 4.18
The newly developed nanocatalyst of Mazda makes use of Pt nanoparticles smaller than 5 nm embedded in the base material. This prevents agglomeration and, at the same time, reduces the amount of precious metal required for catalysis by 70%. Image courtesy of Mazda.

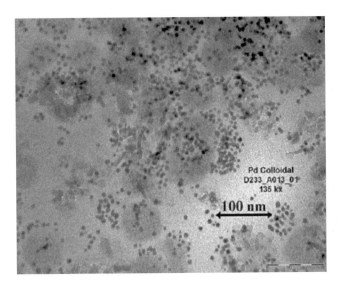

Figure 4.19
The new palladium nanocatalyst series developed by BASF as part of its NanoSelect platform technology (an extension of Mazda's single-nanocatalyst concept) reduces the amount of precious metal required for catalysis by 90% and eliminates the need for toxic lead; another achievement of nanochemistry. Image courtesy of BASF.

Characterized by similar activity and selectivity in specific hydrogenation reactions, these new heterogeneous catalysts are an environmentally compatible alternative to Lindlar catalysts, achieving significant cost savings by using approximately ten-times lower amounts of precious metals (www.catalysts.basf.com).

The nanochemistry technology combines reducing and stabilizing functions to generate highly unimodal, nanosized metal colloids (Figure 4.19).

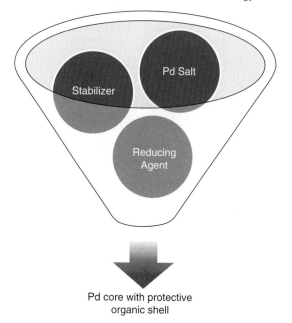

Figure 4.20
The NanoSelect technology is a highly versatile nanochemistry technique affording protected and highly selective catalytic metallites (contrary to traditional, unprotected catalysts shown at the bottom right). Image courtesy of BASF.

These colloids can then be deposited onto different support materials to give heterogeneous catalysts with unique and surprising catalytic behaviors. One such catalyst offers equivalent or greater hydrogenation activity to Lindlar catalyst with no lead content and an order of magnitude less metal (0.5 wt% compared to 5 wt%), thus significantly decreasing the cost of the process and preventing lead contamination of products that are then used as drugs, cosmetics, or food additives.

Once again, the innovation has been made possible by a nanochemistry technique (Figure 4.20) affording narrow metal crystallite size distribution, whereas an organic shell protects the metal particle from either agglomeration or passivation by external reactants.

References

1. LaMattina, J.L. (2008) *Drug Truths: Dispelling the Myths About Pharma R&D*, John Wiley & Sons, Inc., Hoboken.
2. Cue, B.W. Jr. (2005) *Chem. Eng. News*, **83** (39), 46.
3. Anastas, P. and Warner, J.T. (2000) *Green Chemistry : Theory and Practice*, Oxford University Press, New York.
4. Parsons, H. (2008) The Greening of Pharma. PharmaManufacturing.com, http://www.pharmamanufacturing.com/articles/2008/054.html?page=4 (accessed October 30, 2009)
5. Thomas, J.M. (2008) Heterogeneous catalysis: enigmas, illusions, challenges, realities, and emergent strategies of design. *J. Chem. Phys.*, **128**, 182502.
6. Jones, M.D., Raja, R., Thomas, J.M., Johnson, B.F.G., Lewis, D.W., Rouzaud, J., and Harris, K.D.M. (2003) *Angew. Chem. Int. Ed.*, **42**, 4326.
7. Zhang, X., Stefanick, S., and Villani, F.J. (2004) Application of microreactor technology in process development. *Org. Process Res. Dev.*, **8**, 455.
8. Broadwater, S.J., Roth, S.L., Price, K.E., Kobašlija, M., and Tyler McQuade, D. (2005) *Org. Biomol. Chem.*, **3**, 2899.
9. Chandler, A.D. Jr. (2005) *Shaping the Industrial Century: The Remarkable Story of the Evolution of the Modern Chemical and Pharmaceutical Industries*, Harvard University Press.
10. Ciriminna, R. and Pagliaro, M. (2004) *Curr. Org. Chem.*, **8**, 1851.
11. Avnir, D., Levy, D., and Reisfeld, R. (1984) *J. Phys. Chem.*, **88**, 5956.
12. Avnir, D. (1995) *Acc. Chem. Res.*, **28**, 328.
13. Ciriminna, R. and Pagliaro, M. (2006) Recent uses of sol-gel doped catalysts in the fine chemicals and pharmaceutical industry. *Org. Process Res. Dev.*, **10**, 320–326.
14. Ciriminna, R. and Pagliaro, M. (2003) *Chem. Eur. J.*, **9**, 5067.
15. Michaud, A., *et al.* (2007) SiliaCat™ TEMPO: an effective and useful oxidizing catalyst. *Org. Process Res. Dev.*, **11**, 766–768.
16. Abu-Reziq, R., Blum, J., and Avnir, D. (2004) *Chem. Eur. J.*, **10**, 958.
17. Livage, J. (2001) Le mariage du verre et du vivant. *La Recherche*, **342** (1 May). http://www.larecherche.fr/content/impression/article?id=12853.
18. Avnir, D., Ottolenghi, M., Braun, S., and Zusman, R. (1994) US patent 5,292,801.
19. Reetz, M.T., Tielmann, P., Wiesenhofer, W., Konen, W., and Zonta, A. (2003) *Adv. Synth. Catal.*, **345**, 717.
20. Carturan, G., Dal Toso, R., Boninsegna, S., and Dal Monte, R. (2004) *J. Mater. Chem.*, **14**, 2087.
21. Lee, M.-Y., Park, C.B., Dordick, J.S., and Clark, D.S. (2005) Metabolizing enzyme toxicology assay chip (MetaChip) for high-throughput microscale toxicity analyses. *Proc. Natl. Acad. Sci. U.S.A.*, **102**, 983.

5
Organically Doped Metals

5.1
A Watershed Development in Science

In 2002 David Avnir and his team at the Hebrew University of Jerusalem published a research paper [1] in which they introduced a new family of hybrid materials at the border between metals and organic molecules, showing how to entrap several dye molecules within silver (Figure 5.1). Using an approach similar to that of doped sol-gel materials, the authors wrote, "The next "Everest awaiting to be climbed from a similar point of approach has been the doping of metals with organic molecules."

The entrapment leaves the molecules intact and shows that, as in the case of doped sol-gel materials, the entrapped molecules are *accessible* to chemical interaction with an external reagent. The organic dopant is immersed in a sea of electrons, altering its chemistry, and new properties that are not found in *either* of the components emerged. In other words, organic molecules, small or polymeric, are entrapped within metals at molecular level to yield true metal-organic alloys (MORALs).

In the subsequent seven years the team has rapidly developed this new materials technology, showing its great ability to tailor metals – including silver, copper, gold, palladium, and more – with any of the properties of the vast library of organic molecules, with the potential for creating materials that are "in-between" metal and, say, plastics, as they soon also showed how to entrap polymeric molecules [2]. In Avnir's words:

> "One can only imagine the huge potential which can be opened by the ability to tailor to metals with any of the properties of organic molecules."

> "Metals will then have not only the traditional properties and applications, but also many new properties which will merge their classical virtues with the diverse properties of organic molecules."

Nano-Age: How Nanotechnology Changes our Future. Mario Pagliaro
© 2010 WILEY-VCH Verlag GmbH & Co. KGaA, Weinheim
ISBN: 978-3-527-32676-1

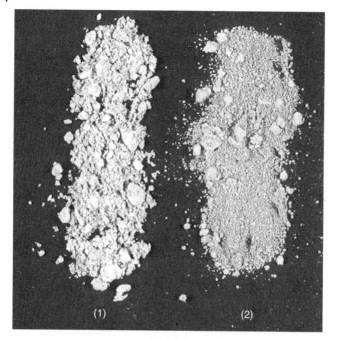

Figure 5.1
Metal silver in powder form (1) prepared according to Scheme 5.1 can be easily doped with an organic dye to yield (2) a metal-organic alloy (MORAL). Adapted from Reference [1], with permission.

The work simply had no prior art to cite as these artificial materials had previously been unknown. A watershed in chemistry had been reached almost silently!

In general, these materials are expected to have an impact wherever metals are used: catalysis, electrochemistry, magnetism, corrosion protection, and the classical uses of metals as materials. For instance, one can induce unorthodox properties, such as producing silver that acts as an acid, or having right- and left-handed enantiomers of chiral gold and silver, or forming superior catalysts, and so on.

The entrapment methodology proved versatile for different types of molecules, including hydrophilic and hydrophobic small molecules and polymers, as well as bioactive molecules. Two principal methods were developed for the synthesis of these composites, both of which are based on reduction of the metal-cation in the presence of the molecule to be entrapped (the "dopant").

In one method, the process is homogeneous, in that all components, including the reducing agent, are soluble (Scheme 5.1). In the second method [3], the metal cation is reduced by a heterogeneous dispersion of a sacrificial reducing metal. In general, these composites are porous: detailed

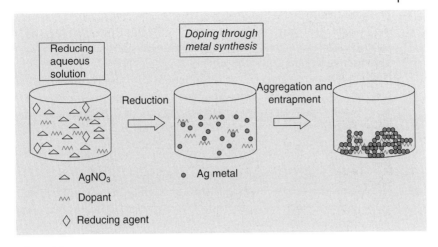

Scheme 5.1
In one of the two general methods that afford MORALs, doping of the metal is simply carried out by synthesizing the metal by mild reduction of the metal ions in solution in the presence of the organic dopant. Image courtesy of David Avnir.

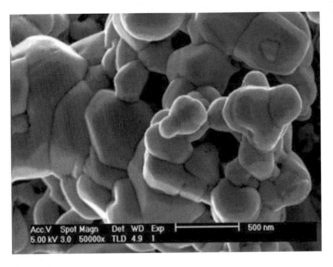

Figure 5.2
A microphotograph reveals the crystallite aggregated nature of the MORALs material; shown here is the structure of silver doped with poly(styrene sulfonic acid). Reproduced from Reference [2], with permission.

studies provide a structural picture of physical caging inside partially closed pores, the walls of which are the faces of nanocrystallites (Figure 5.2).

These new materials, metallo-organics, are the opposite of organometallics. The possibilities of imprinting metals with organic molecules are practically endless, and the process itself is so simple that it can be used in high-school demonstration experiments. This entrapment process "is a watershed development in science," commented organometallic chemist Howard Alper of the University of Ottawa cited in [4].

When, however, scientists applied for grants to explore the feasibility of their idea, the project was rejected, showing once again how important it is for funding agencies in the nanotechnology sector to finance new ideas, especially those with a very fundamental scope such as in the case of organically doped metals. One point deserves to be emphasized:

> "The project was made possible mainly by using the existing infrastructure, since funding agencies traditionally do not support projects which have a flavor of 'I have an idea which apparently has not been tested before and, furthermore, I do not know if it will work, but anyway please provide me with the means to check it'."
>
> "This situation is quite familiar, we believe, with many of our readers [1]."

The scientists entered that challenge "because it was there," refusing to answer questions on applications (which became plentiful in subsequent reports).

5.2
The New Reactivity of Metal-Entrapped Molecules

In MORALs the dopant is accessible for reaction by substrates diffusing into the composite; despite being accessible, the dopant does not leach out (even in solvents it is normally soluble in), or leaches out to a negligible extent. The interactions of the entrapped molecules with the metallic matrix account for the non-leachable nature of the entrapped dopants, being based on multiple physical and chemical interactions through the typical moieties of the organic molecules, (such as $-NH_3$, $-CO_2$, $-SH$, etc., which are known to have affinity for metallic surfaces). Overall, this opens up a whole new field of chemical *reactivity*, greatly expanding the chemistry possibilities.

In addition to the possible induction of unorthodox properties listed in Section 5.1, a metallic matrix can also serve as a heterogeneous carrier for enzymes. For instance, the acidic enzyme phosphatase was recently entrapped in gold and silver, by room temperature reduction of the metal cation in the presence of the enzyme [5].

The entrapped enzyme (Figure 5.3) is accessible for biochemical transformation interaction with a diffusing substrate, and the matrix protects efficiently against extreme pH conditions, as shown by the fact that the acidic enzyme is kept alive under basic conditions.

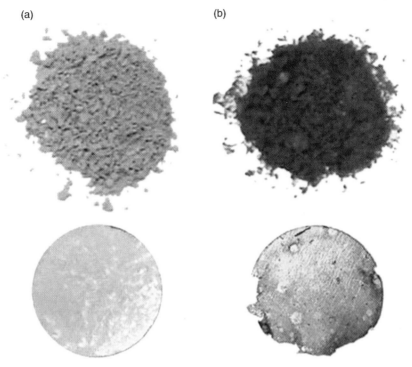

Figure 5.3
Enzymatically active metallic composites, powders, and pressed coins: (a) AcP@Ag and (b) AcP@Au. Reproduced from Reference [5], with permission.

Entrapment of enzymes within metals is very different from adsorption of enzymes on pre-aggregated metals, in both conceptual and reactivity aspects. While doping of the metal results in a dispersion that mimics a 3D molecular distribution – the entrapped enzyme "sees" metal all around (Figure 5.4) – adsorption, in contrast, provides a 2D arrangement and thus the adsorbed enzyme molecules can aggregate and block their own activity.

Another useful application is the induction of new, unorthodox reactivities to the metals, as shown for instance by the making of acidic or basic silver [6]. Basic silver is obtained by entrapment of the basic polymer PVBA [poly(vinylbenzyltrimethylammonium hydroxide)], affording a highly reactive material, PVBA@Ag, suitable, for example, for ion-exchanging electrodes (Figure 5.5).

Acidic silver is instead obtained by entrapment of the acidic polymer Nafion, affording a reactive acidic solid that is highly active in catalytic reactions requiring a strong acid (Figure 5.6). In addition, a fundamental physical property of a metal such as conductivity can be largely altered by the metallo-organics approach.

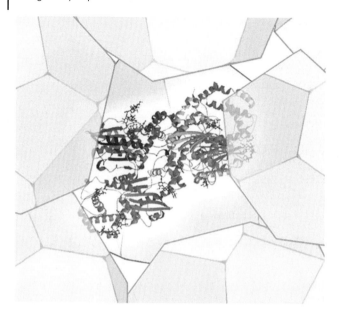

Figure 5.4
Schematic view of entrapped enzyme within a metal. Reproduced from Reference [5], with permission.

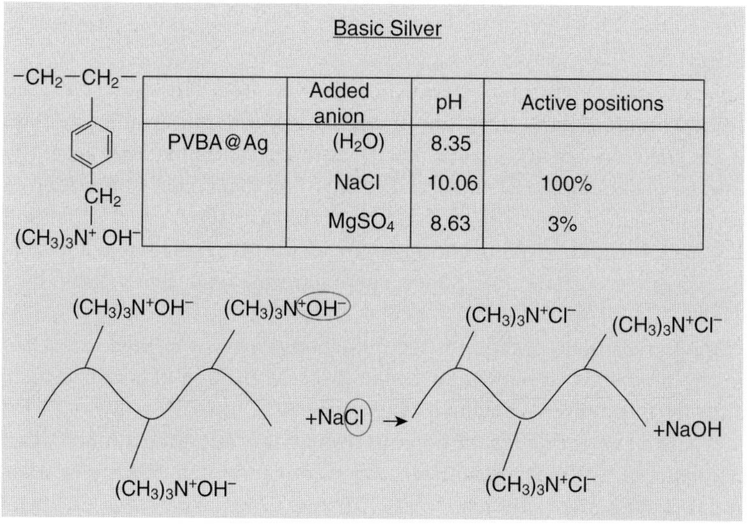

Basic Silver
–CH2–CH2–
CH2
(CH3)3N+ OH–
Added anion
pH
Active positions
PVBA@Ag
(H2O)
8.35
NaCl
10.06
100%
MgSO4
8.63
3%
(CH3)3N+OH–
(CH3)3N+OH–
(CH3)3N+OH–
+NaCl
(CH3)3N+Cl–
(CH3)3N+Cl–
(CH3)3N+Cl–
+NaOH

Figure 5.5
The metal-organic alloy made of the polymer PVBA entrapped in silver behaves as basic silver, rapidly exchanging its hydroxide ions with chloride ions in solution. Reproduced from Reference [6], with permission.

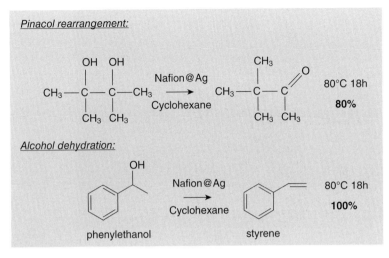

Figure 5.6
The metal-organic alloy made of the polymer Nafion entrapped in silver behaves as acidic silver, suitable for several acid-catalyzed reactions. Reproduced from Reference [6], with permission.

For example, the electrical conductivity of pressed films made of polyaniline (PANI) in two molecular weights entrapped in silver at different concentrations differs by order of magnitudes, extending the functional applications of the new metallic composites and offering insight into the polymer–metal molecular level interactions [7].

This time, however, the catalyst is easily removed from the reaction mixture at the end of the reaction, whereas its pronounced reactivity makes it an interesting option for several industrial processes in which acid-catalyzed reactions are used.

5.3
Two-for-One-Catalyst

By entrapping a soluble rhodium complex inside a porous matrix of silver atoms, Avnir's team at the Hebrew University of Jerusalem initiated a new approach to heterogeneous catalysis [8]. Chemists often harness soluble homogeneous catalysts for surface-based heterogeneous reactions by fixing the catalyst on an organic polymer or an inorganic oxide such as silica, materials that are non-conducting. Trapping the rhodium complex in a "sea of electrons" enhances its catalytic activity, presumably by altering its electron donor–acceptor properties and modifying substrate binding.

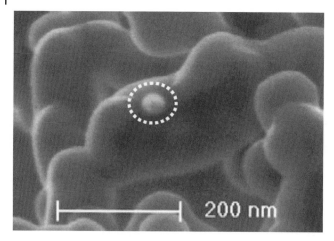

Figure 5.7
Microphotograph of the [Rh]@Ag MORAL, showing the catalytic Rh centers entrapped within the silver nanocrystallites. Reproduced from Reference [8], with permission.

To make the material (Figure 5.7), the researchers added zinc powder to a solution containing a rhodium phosphine cyclooctadiene catalyst and silver nitrate. The zinc reduces the silver cations, and as silver crystallites form they aggregate and precipitate out of the solution, taking the catalyst with it, thus creating a new type of catalytic material: [Rh]@Ag. Several aspects were demonstrated with the development of this heterogeneous catalyst: a metal can be used as a support for heterogenizing a homogeneous catalyst; the homogeneous catalyst is stabilized by entrapment within the metal; the products of the composite catalyst are *different* compared to those obtained from the homogeneous one. Once again, the adsorption of [Rh] on the surface of Ag and its entrapment are very different processes and only the encapsulation provided appreciable catalytic activity.

Thus, while homogeneous [Rh] was destroyed entirely after converting styrene into ethylbenzene in 50% yield, [Rh]@Ag remained active after effecting the same reaction with a yield of 85% (compared to only 7% for [Rh] adsorbed on Ag); in addition, while homogeneous [Rh] hydrogenated diphenylacetylene to bibenzyl (and was completely deactivated after one cycle) with no trace of *cis*-stilbene, [Rh]@Ag afforded *cis*-stilbene as the main product. Moreover, the catalyst could be reused.

The metallo-organics approach to catalysis is not limited to the highly selective processes typical of the low-volume reaction of fine chemistry such as the hydrogenation mentioned above. For example, the selective oxidation of methanol to formaldehyde with oxygen, a multi-million ton reaction carried out in industry using silver as catalyst, turns out to be greatly enhanced when using a metallo-organics silver catalyst [9].

Organic doping of the metal silver with Congo Red (CR@Ag) improves the performance of Ag as a catalyst for methanol oxidation to formaldehyde,

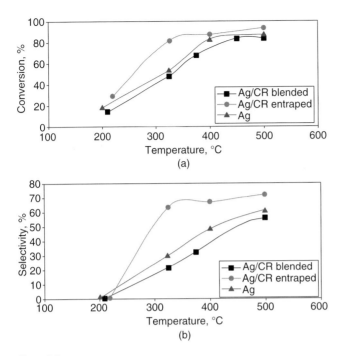

Figure 5.8
In the aerobic oxidation of methanol to formaldehyde, the use of the MORAL made of the dye Congo Red entrapped in silver, CR@Ag, reduces by 100 °C the temperature needed to reach maximum conversion (a), and reduces by 200 °C the temperature at which maximal selectivity is obtained (b). Reproduced from Reference [9], with permission.

outperforming both pure Ag and CR-coated Ag (CR/Ag) in terms of lowering by 100 °C the temperature needed for maximal conversion, lowering by 200 °C the temperature required to reach the maximal selectivity (aldehyde formation), and increasing the maximal space velocity by a factor of two (Figure 5.8). The scientists were led to this discovery by a detailed investigation of the thermal behavior of CR@Ag under an oxidative atmosphere, which indicated that the metal strongly catalyzes the CR oxidation, pointing to the relevant temperature for activation of the catalyst.

5.4
Chiral Metals

Not only new physical and chemical properties, but also new optical properties of metals become accessible by entrapment, for example, of a chiral dopant within the cages of metal nanocrystallites. This was achieved by encapsulation of the different enantiomers of two chiral molecules. Like all metals, MORALs also exhibit the photoelectric effect: when a high-energy photon hits the metal an electron is ejected.

However, the chiral gold coins ejected different electrons when exposed to clockwise polarized or anticlockwise polarized photons, proving the metal's chirality (Figure 5.9) [10].

Chirality is a Greek term that means that an object, for example a molecule, has a *mirror image* (in the same manner as left and right hands).

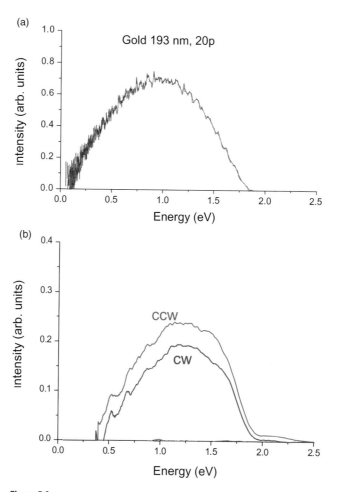

Figure 5.9
Whereas the scattering of both left and right components of polarized light from undoped, bulk gold is identical (a), scattering from gold doped with one enantiomer of an organic molecule (L-quinine, b) is completely different. Reproduced from Reference [10] with permission.

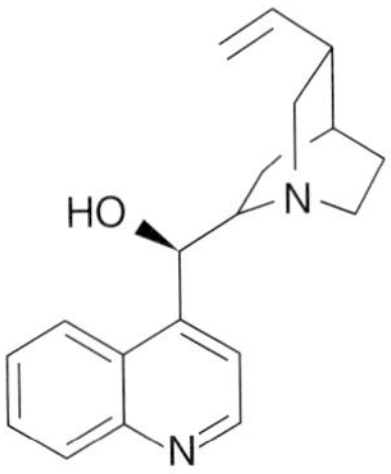

Cinchonidine

Figure 5.10
The alkaloid cinchonidine, the chiral molecule that was used to imprint and prepare the first catalytic chiral palladium sample.

Chiral molecules have asymmetrical centers (right- or left-handed structures). Metals are not chiral because they have neither.

What is chiral in the doped metal are the chiral dopants that distort chirally the molecular orbitals of the metal, as well as the geometry of the metal cages around the dopant molecules.

The next step, therefore, was the development of a new type of catalyst, namely, chiral metals. In well-established organometallic inorganic chemistry it took some 100 years before organometallics gave us chiral catalysts, but the nanochemistry approach gave us chiral catalysts in less than two years when a Spanish PhD student in catalysis working at Gadi Rothenberg's laboratory at the University of Amsterdam made a new chiral palladium metal catalyst by using organic template molecules that were then later removed.

Using simple precipitation technology, the team managed to imprint palladium metal crystals with a chiral organic template. In particular, they reduced a palladium salt in the presence of cinchonidine or closely related chiral alkaloids (Figure 5.10) [11].

Again, on the basis of spectroscopy measurements made with circularly polarized light, the team demonstrated that the composites have imprinted chirality and that extracting the organic fraction with solvents forms porous metal products that are also chiral.

Dissolving the alkaloid leaves pure palladium – a black powder – which is imprinted with the enantiomer it has lost, that is, a chiral cavity is left in the palladium metal. The metal itself retains all its usual properties, such as malleability, conductivity, and catalytic activity. Moreover, by using a ten-tonne French press, the researchers even pressed a chiral palladium coin, roughly the size of a two-cent piece (Figure 5.11).

The team tested samples of chiral palladium as catalysts for the hydrogenation of isophorone, proving that they could produce modest excesses

(1) (2) (3)

Figure 5.11
Chiral palladium powder (3) pressed into a coin (2), with a real coin for scale (1). Photograph courtesy of Paul Collignon, University of Amsterdam.

of left- and right-handed enantiomers. For example, they showed that hydrogenation of isophorone, a cyclic keto-olefin, yields the *(R)*-enantiomer of the corresponding cyclohexanone with 16% enantiomeric excess. The reported catalytic performance is modest, but clearly this strategy holds promise for developing robust heterogeneous asymmetric catalyst as there is great scope for both fundamental and applied extensions to this work.

An attractive feature of the nanomaterial, for example, is that the chiral metal surface appears to have a high surface area to weight ratio, making it attractive for large-scale industrial processes. Palladium is still a very expensive material to use pure as a catalyst and enantiomeric excesses far greater than those reported here will be required before the process becomes financially viable. Other applications might be that of chiral sensing working via a simple "lock and key" mechanism – one enantiomer would fit the metal's chiral pores better than another.

References

1 Behar-Levy, H. and Avnir, D. (2002) Entrapment of organic molecules within metals: dyes in silver. *Chem. Mater.*, **14**, 1736–1741.

2 Behar-Levy, H., Shter, G.E., Grader, G.S., and Avnir, D. (2004) Entrapment of organic molecules within metals, II: polymers in silver. *Chem. Mater.*, **16**, 3197–3202.

3 Yosef, I. and Avnir, D. (2006) Metal-organic composites: the heterogeneous organic doping of the coin metals-copper, silver, and gold. *Chem. Mater.*, **18**, 5890–5896.

4 Ritter, S.K. (2008) *Chem. Eng. News*, **86** (35), 9.

5 Ben-Knaz, R. and Avnir, D. (2009) Bioactive enzyme-metal composites: the entrapment of acid phosphatase within gold and silver. *Biomaterials*, **30**, 1263–1267.

6 Behar-Levy, H. and Avnir, D. (2005) Silver doped with acidic/basic polymers: novel, reactive metallic composites. *Adv. Funct. Mater.*, **15**, 1141–1146.

7 Nesher, G., Aylien, M., Sandaki, G., Avnir, D., and Marom, G. (2009)

Polyaniline entrapped in silver: structural properties and electrical conductivity. *Adv. Funct. Mater.*, **19**, 1293–1298.

8 Yosef, I., Abu-Reziq, R., and Avnir, D. (2008) The entrapment of an organometallic complex within a metal: a new concept for heterogeneous catalysis. *J. Am. Chem. Soc.*, **130**, 11880–11882.

9 Shter, G.E., Behar-Levy, H., Gellman, V., Grader, G.S., and Avnir, D. (2007) Organically doped metals – a new approach to metal catalysis: enhanced silver-catalyzed oxidation of methanol. *Adv. Funct. Mater.*, **17**, 913–918.

10 Behar-Levy, H., Neumann, O., Naaman, R., and Avnir, D. (2007) Chirality induction in bulk gold and silver. *Adv. Mater.*, **19**, 1207–1211.

11 Pachon, L.D., Yosef, I., Markus, T., Naaman, R., Avnir, D., and Rothenberg, G. (2009) Chiral imprinting of palladium with Cinchona alkaloids. *Nat. Chem.*, **1**, 160–164.

6
Protecting Our Goods and Conserving Energy

6.1
Multifunctional Nanocoatings

Acting as physical and chemical protection agents and providing several brand new properties to the coated substrate, functional coatings are immensely important to industry and society in general. Accordingly, a multibillion dollar industry exists that historically has relied upon organic polymers and – to a far lesser extent – also upon the employment of inorganic polymers such as silica or alumina (www.specialchem4coatings.com/).

In general, in addition to innovations on the energy supply side, achieving a sustainable energy future will also require much more efficient energy and raw materials use and conservation (Figure 6.1).

Materials and energy conservation technologies will have a major impact, especially for economically developed countries that are major energy consumers of goods as well as of energy. These conservation measures include the development of advanced insulating materials, adaptation of new lighting technologies such as solid-state OLED lighting, and multifunctional surface enhancement for all sorts of goods. Solar energy conversion, energy storage and conservation, and surface protection can all directly benefit from nanoscale design.

For example, the reason water, dirt, and stains remain on many surfaces is because the pores on the materials surface are large and spread apart. Nanostructured coatings penetrate into the pores of industrially relevant surfaces such as those of cars, boats, aircrafts, and buildings, providing long-term protection against dirt, water stains, mold, mildew, chipping, and scratching.

In the USA, the company Diamon-Fusion (www.diamonfusion.com) has commercialized a powerful nanochemical technology for coating with multifunctional characteristics, including water and oil repellency (hydrophobic and oleophobic), impact and scratch resistance, protection against graffiti, dirt, and stains, UV stability, additional electrical insulation, protection

Nano-Age: How Nanotechnology Changes our Future. Mario Pagliaro
© 2010 WILEY-VCH Verlag GmbH & Co. KGaA, Weinheim
ISBN: 978-3-527-32676-1

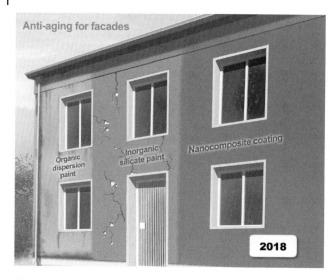

Figure 6.1
The new nanobinder Col.9 combines in one product the different advantages of conventional coating types in terms of low dirt pick-up, chalking and rack resistance, and color retention. Reproduced from www.col9.com, with permission.

against calcium and sodium deposits, and increased brilliance and lubricity. The nanocoating (Figure 6.2) works at nanoscale levels to change the molecular composition of any silica-based surface such as glass, ceramic tile, porcelain, and granite. This provides, in chemical terms, an optically clear nanofilm along with a strong and durable covalent bond. The technology uses a chemical vapor deposition process. The vapors first react with the moisture on the surface and the silica in the substrate to be treated. In stage 1 a chemical reaction causes a "crosslinked" silicone film to be grown from below the surface out. After converting the chlorine atoms into OH groups using additional moisture (chlorine was left at the end of the atom chains after the first stage), a second specially formulated vapor is introduced to the surface. The second stage "caps" the entire chain of atoms. This unique "capping" substantially increases the hydrophobicity and durability, leaving, chemically speaking, no points of attachment for contaminants and creating a truly repellant charge.

The chemical reaction bonds to form an ultrathin protective layer of optically clear durable material, a nanostructured device, making the surface significantly easier to clean and more resistant to weathering. The bond created is a covalent bond, meaning that the coating shares the electrons within the glass itself, thus becoming a part of the glass. Covalent bonds are approximately ten times stronger than hydrogen-bridge bonds, which are commonly present in most other water-repellent coatings.

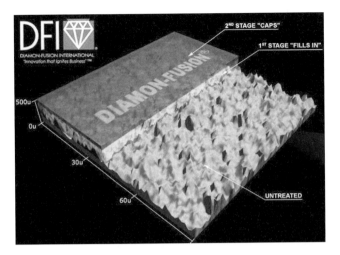

Figure 6.2
DFI nanotechnology works in two steps. In stage 1 a chemical reaction causes a "crosslinked" and "branched" silicone film to be grown from below the surface outwards. The second stage "caps" the entire chain of atoms, increasing the hydrophobicity and durability and leaving, chemically speaking, no points of attachment for contaminants, thereby creating a truly repellant charge. Reproduced from diamonfusion.com, with permission.

Applications of the coating to glass span all main sectors in which glass is a construction component, from cars to windows, skylights, screens, lenses, and so on. With such a coated outdoor glass, for example, rain no longer causes the formation of mud trails as water and dirt accumulated on the surface run off, leaving the surface clean and intact.

A general principle of nanoscience is that the smaller the particle the closer they reside together, which alters the surface area and material characteristics dramatically, affording glass or metal particles that exhibit very high strength, hardness, corrosion resistance, and even transparency in coatings. Thus, another American company, Nanovere Technologies, has commercialized a multifunctional nanocoating that can prevent surface scratching, reduce surface cleaning and maintenance, and reduce barnacle adhesion on boat hulls and water drag resistance. A complete line of nanocoating formulations named Zyvere is available to the automotive, aerospace, marine, and industrial markets (www.nanocoatings.com). The active nanoparticles are typically 7 nm in diameter and show a nanostructured polymer architecture that is responsible for the multifunctional performance. The polymer is the well-known polyurethane but it is chemically altered with many hydroxyl functions, forming what is known as a dendrimer. Dissolved in a solvent the polymer can now encapsulate a plurality of metal oxide nanoparticles; a mild treatment due to ambient moisture is enough to promote condensation of the nanoparticles to form a coating

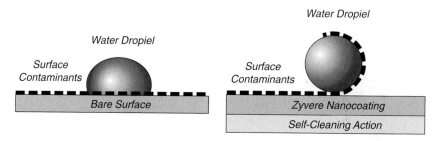

Figure 6.3
Water droplets roll across a cured Zyvere surface picking-up dirt and contaminants and leaving behind a cleaner surface. Reproduced from www.nanocoatings.com, with permission.

that exhibits high strength, hardness, corrosion resistance, and permanent water, oil, ice, and barnacle repellency (hydrophobic effect) (Figure 6.3).

Invented by Thomas Choate, a microbiology graduate of the University of Michigan [1], Zyvere is not a polyurethane clear coating with nanoparticles added to improve the physical properties of the coating, that is, it is not a nano-dust. Rather, all of the physical properties of the coating are embedded into the molecular backbone of the resin.

Another major class of emerging nanostructured multifunctional coatings makes use of organosilica hybrids that are finding applications in fields ranging from the building industry through aircraft and cultural heritage protection. These materials have emerged in the last few years as a broad class of new covering materials capable of serving both the traditional requirements of the industry, namely the versatility of organic polymers, along with the strength and durability typical of inorganic polymers.

Such coatings, especially in the form of water-based formulations that abate the VOC (volatile organic compound) content of traditional coating paints, are of special interest since their properties—intermediate between those of polymers and glasses—can deliver specific and unique requirements not obtainable by organic polymers and glasses alone. For instance, the facade of a building painted with one of these new paints containing the sol-gel hybrid formulation (Figure 6.1) recently commercialized by BASF will now retain its fresh aspect for years. Trade-marked Col.9® (www.col9.com) 50% of this product consists of a dispersion of organic plastic polymer particles in which nanoscale particles of silica are incorporated and evenly distributed. Thanks to this combination of elastic organic material and hard mineral, coatings based on this novel nanobinder combine in one product the different advantages of conventional coating types in terms of low dirt pick-up, chalking and crack resistance, and color retention (Figure 6.4).

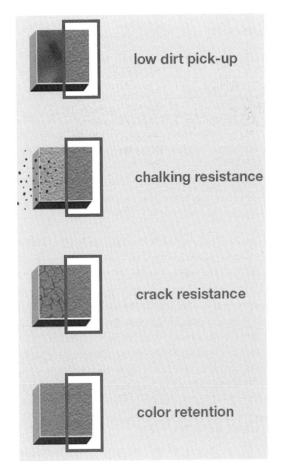

Figure 6.4
Color retention of colored surfaces treated with the nanobinder Col.9. Reproduced from www.col9.com, with permission.

The inorganic nanoparticles are homogeneously embedded in larger acrylate polymer particles. After drying and curing at room temperature, dirt repellent nanostructures form on the surface of the coating, reinforcing the hydrophilic properties of the facade. The molecular composite nature of the particles ensures that the nanoparticles will remain homogeneously fixed instead of agglomerating when the liquid coating is formulated with water and color pigments; this enables the formation of a stable three-dimensional network of nanoparticles that covers the entire surface of the film. Because of its high silica content, however, the nanocomposite of Col.9 does not have thermoplastic tackiness. At the same time, the mineral

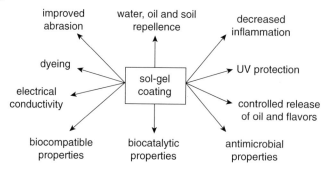

Figure 6.5
Some possibilities of surface functionalization by modified nanosols.

particles provide the coating with a hydrophilic attracting surface on which rain drops are immediately dispersed. As regards cleanliness, this offers a dual benefit: in heavy rain, particles of dirt are washed off extensively from the facade surface. In addition, the thin film of water remaining when rain has stopped dries extremely quickly, which prevents mold formation. In contrast, the rain rolling off unevenly in thick droplets from water-repellent surfaces of fully synthetic resin coatings often leaves behind unattractive streaks of dirt. Finally, permanently incorporating the mineral nanoparticles in the much larger acrylate polymer particles means that the color tone remains stable and there is no surface chalking even after years of exposure to weather.

In general, sol-gel coating thus enables the manifold alteration of the physico-mechanical, optical, electrical, and biological properties of a material's interface (Figure 6.5). Depositing a sol-gel silica-based coating on a piece of substrate is easy and economically attractive as it changes and largely enhances the physical and chemical properties of the coated substrate while consuming limited amounts of valued silicon-based liquids used in the manufacture. At small film thickness, for example, sol-gel coatings feature excellent scratch, scrub, and weathering resistance along with superior thermal stability. Alternative coating methods such as chemical vapor deposition are generally energy- and capital-intensive, requiring the use of dedicated plants and deposition processes.

Silica and organosilica coatings, instead, are obtained under very mild conditions in the liquid phase by the so-called chemical sol-gel process [2]. Explored since the early 1980s this method of creating new glassy functional materials at room temperature, starting from a liquid, allows the mixing of inorganic and organic components at the nanometric scale to yield highly versatile functional materials at the interface of organic and inorganic realms; the chemical and physical properties of these materials can thus be tailored to meet the demanding requirements of widely different chemical applications [3].

Numerous ORMOSIL-based coating materials are commercially employed as protecting glasses against scratch and abrasion for sensitive surfaces. Coating formulations are generally clear, transparent, low viscosity, solvent-free liquids that are stable with a shelf life of several months and are supplied by a large number of companies worldwide, such as, for instance, Evonik in Germany and Safe Marine Nanotechnologies in Italy.

Both companies have recently commercialized a new sol-gel product line trademarked Dynasylan® (www.dynasylan.com), which provides good corrosion and abrasion protection for valuable metals, and also provide excellent adhesion of coating. Despite its acute toxicity carcinogenic Cr(VI), unfortunately, remains an essential ingredient used by industry for corrosion control. For example, contamination of drinking water with hexavalent chromium in Hinkley, California was due to industrial use of hexavalent chromium to fight corrosion in a cooling tower. Some of the wastewater percolated into the groundwater, causing several illness cases.

Clearly, its replacement with doped hybrid sol-gel coatings offers also an enormous environmental benefit. In general, the application of sol-gel as a means of corrosion inhibition is based on two complementary functions: the physical barrier of the polymeric matrix and the specific corrosion inhibition of the organic moiety in the ORMOSIL.

Established by a team of Italian and British researchers [3], Safe Marine Nanotechnologies (www.safenanotech.com) commercializes an anticorrosive paint for yachts, bridges, and ships, based on silica hybrids capable of high abrasion resistance and low thermal expansion due to the presence of nanostructured bicontinuous domains. The paint is made of epoxy-silica hybrids doped with a metal ion in which the two components – silica and the epoxy resin – behave synergistically, with the inorganic component enhancing the network density and the organic group easing formation of the bicontinuous domain. The molybdenum metal ions are continuously and slowly released to prevent (inhibit) corrosion. The result is a nanomaterials-based paint called Ballast SAFE (Figure 6.6) with unique characteristics: high solids content (96%) and solvent-less, which upon application provides high adhesion (>25 MPa) and high gloss (>90 gloss units).

Sol-gel protection of metal surfaces is being explored and will soon lead to new results of large societal and economic impact. Magnesium, for example, has high electrical and thermal conductivity, is abundant and easily recycled, and its high strength to weight ratio makes it a valuable asset in the transportation and aviation industry. However, its utilization has not reached full capacity because of its high chemical reactivity and tendency to corrode. The native oxide layer does not prevent its pitting corrosion, facilitated mostly by halides, which results in destructive effects. The team of Daniel Mandler at the Hebrew University of Jerusalem is to use a sol-gel coating for the controlled corrosion of what is called

Figure 6.6
Microphotograph (a) of the epoxy-silica hybrid coating acting as active corrosion inhibitor following application of the Ballast SAFE paint to large metal vessels (b). Reproduced from safenanotech.com, with permission.

biodegradable magnesium. The idea, which is already commercialized, is to use Mg for medical implants that slowly dissolve and therefore do not require a second operation to remove them. The problem with these implants is their uncontrolled corrosion, which the team hopes to solve by the sol-gel methodology.

Hydrophobic silane formulations such as those of water-based organosilanes are good surface protection agents for wood as the resulting coating repels *both* oil (dirt) and water, protects the surface from the effects of weathering, and reduces the growth of microorganisms. In general, the alkyl groups ensure durability and provide hydrophobicity, whereas oleophobicity is due to the siloxane backbone structure. The nanoparticles penetrate the wood's pores, acting as wood reinforcement and at the same time the modified silica coating imparts flame retardant, water repellent, and antimicrobial properties.

Even more importantly from an environmental viewpoint is the new environmentally friendly, mechanically stable, and inexpensive fouling-resistant and fouling-releasing coating AquaFast [4]. Developed by US scientists Mike Detty and Frank Bright at Buffalo, The State University of New York, this hybrid formulation is highly effective for both fast-moving ships and ships in dock. Its functioning is simply based on the hydrophobicity impatred to the immersed surface by an ORMOSIL coating. The extremely low critical surface tension of the coated surface inhibits settlement of zoospores of marine fouling algae as well as of juveniles of tropical barnacles. Figure 6.7 shows the real-world performance of AquaFast.

Once again, it is the versatility of the nanochemistry approach to materials synthesis that by enabling the easy incorporation of organic functional groups in the hybrid material provide surfaces of different wettability and critical surface tension.

27' Bayliner as pulled from water November 7, 2005 (small white area wiped clean with a "finger"!)	27' Bayliner after sitting in the sun one day The alga is sloughing off to give a clean surface – no friction or washing required!
(a)	(b)

Figure 6.7
Pictures of a boat moored in Irondequoit Bay (Lake Ontario), Rochester, NY. Image courtesy of Professor Frank V. Bright.

6.2 Multifunctional Textiles

The maintenance and improvement of current properties and the creation of new material properties are important reasons for the functionalization of textiles with sol-gel organosilica coatings with particle diameters smaller than 50 nm (nanosols) [5]. Within today's global textile market, worth more than $400 billion, high-grade textiles rapidly grow as functional textiles improve applications (enhanced comfort, easy care, health, and hygiene) and extend their range of utilization (ensuring protection against mechanical, thermal, chemical, and biological attacks), affording technical textiles with numerous usages in automotive, railroad, and aviation engineering, in construction, and in the home.

New products that are being developed include textiles with water, oil, and soil repellency and with antimicrobial properties. The nanoparticulate size of the sol particles promotes excellent adhesion to the textile fibers, which can be further enhanced by subsequent thermal treatment (Figure 6.8). For example, sol-gel immobilized bioactive liquids such as cineol, camphor, menthol, evening primrose, and perilla oil have been used to functionalize textiles to afford either skin-friendly textiles with antimicrobial and antiallergy effects due to immobilized natural oils or textiles for therapeutic treatment of the respiratory tract by means of immobilized mixtures of high volatility natural agents such as eucalyptol, camphor, and

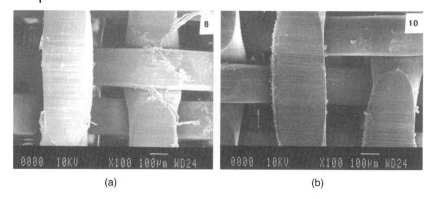

Figure 6.8
SEM pictures of polyester sieves after abrasion: (a) without and (b) with nanosol coating. Reproduced from Reference [5], with permission.

menthol [6]. Owing to the very low layer thickness (below 1 μm) the consumption of coating solutions is also very low; in addition, the use of coated textiles offers a more continuous and prolonged release of the embedded liquid, which is also more convenient. The first optimized commercial sol-gel products for textile refinement were manufactured by CHT (Tübingen, Germany), which sells several water-based (non-inflammable) sols for application onto textiles, resulting, for example, in increased stiffness of textiles [7].

6.3
Protecting Cultural Heritage

Sol-gel nanotechnology is now widely applied in the conservation of art objects and cultural heritage. For example, a hybrid silica coating now protects the fourteenth-century mosaic situated above a gate of St Vitus cathedral, in the center of Prague Castle [8]. For centuries, corrosion had obscured the more than 1 million colored tiles and gilt under an opaque, white-gray layer of decay. Earlier restoration attempts did not work. Shortly after each cleaning and preservation, the mosaic would return to its previous grungy state. Eric Bescher, a scientist at the University of California at Los Angeles thus designed a coating for treatment of the entire mosaic based on a multilayer system in which an organic–inorganic sol-gel layer is placed between the glass substrate and a fluoropolymer coating (Figure 6.9).

The resulting coating is flexible, as the mosaic expands and contracts with temperature changes, and is reversible to allow the coating to be removed

Figure 6.9
The sol-gel protective layer on top of the fourteenth century mosaic known as The Last Judgment, above a gate of St Vitus cathedral in the third courtyard of Prague Castle, was applied by simple brushing. Reproduced with permission of the Getty Conservation Institute.

if it does not work well. The sol-gel layers underneath the top polymer layer are estimated to last for 25 years. Beyond the long-term stability under severe aging conditions compared to polymers, the advantage of such materials lies in their ability to adhere to many substrate, the possibility of making thicker coatings than with purely inorganic sol-gel systems, and the ease of application using a brush, by hand, to avoid coating the interstitial space between tesserae (Figure 6.9), followed by curing at 90 °C for two hours using large infrared lamps.

Remarkably, all organic polymers used in previous protection attempts have *failed* to stop the corrosion, because of their poor durability, poor adhesion to the glass, and large diffusion coefficients for SO_2 and water. The mosaic indeed is a large outdoors panel, 13 m wide and 10 m high, made of about 1 million pieces of multicolored, high-potassium glass embedded in a mortar. The glass is chemically unstable as it is exposed to harsh weather conditions (high SO_2 levels, rain, and temperature varying between −30 °C and +65 °C), under which the alkali reacts with the atmosphere and water to form salt deposits.

6.4
Protecting Goods from Light

Ultraviolet (UV) light, either natural or artificial, causes organic compounds to decompose and degrade, because the energy of the photons in UV light is high enough to break chemical bonds. Organic materials, such as polymers, paints, pigments, and dyes, are used in everything from car parts to fine art. Polymers exposed to UV light can lose mechanical strength and integrity, while UV light causes the cellulose and lignin in wood to degrade, discoloring the wood and eventually causing fractures and cracking. The dyes in paintings and photographs progressively fade under UV light and paper becomes yellowed and brittle. UV light is the main factor responsible for the degradation of wooden furniture, plastic parts used in the car industry, and artwork in museums, which are all exposed to natural or artificial lighting for long periods of time.

All this makes protecting light-sensitive materials against UV irradiation an important technological demand in almost every industrial field. One of the most widely used methods of UV protection is the dispersion of UV-absorbing molecules into a material. UV-absorbers must be colorless (or nearly colorless) compounds that show good photostability and can transform the absorbed UV energy into less harmful energy before it reaches the substrate. Inorganic materials, based mainly on mixed metal-oxide films or particles, or organic molecules, such as phenolic molecules, can be used to absorb or scatter light. A well-known example of an inorganic UV-protector is titania (TiO_2), which is commonly used in sunscreens. Coating photosensitive materials with UV-absorbing inorganic or polymer-based films has been studied extensively. Inorganic coatings, however, can only be used on heat resistant substrates, due to the high curing temperature required for their preparation. In addition, polymer films have proven to have low photostability – the film itself degrading after prolonged irradiation with UV light.

Recently developed UV-protective coatings are based on a dispersion of UV-absorber molecules in a hybrid organic–inorganic sol-gel matrix. The sol-gel method allows the preparation of transparent, solid, and porous inorganic material at low temperatures, and the incorporation of large amounts of organic UV-absorbing molecules in its pores, giving excellent UV protection. These coatings are just 1 μm thick, but can reduce the UV light reaching the substrate to less than 7% of the incident light. The coatings are also highly stable upon prolonged exposure to UV light and are fully transparent in the visible region of the spectrum. This means they can be used to coat a wide range of materials, without affecting the way they look (Figure 6.10). The lifetime of photosensitive materials can be increased from the scale of months to years, making the protective coatings very attractive indeed for commercial applications.

ORMOSIL-based coatings made of *large* amounts of organic UV absorber molecules entrapped in modified silica matrices can reduce drastically the UV light reaching the substrate (Figure 6.1), and hence its photodegradation upon prolonged exposure to UV sources [9]. For example, a thin film of organosilica doped with Rhodamine dye reduces the UV light reaching the substrate to less than 7% of the incident light (Figure 6.11), whereas the degradation of 20% of the dye molecules is 14 times slower in coated samples, making such protective coating very attractive for commercial applications.

Photodegradation of the dye molecules in the coated samples was much slower than with the uncoated samples at 25 °C. The UV-absorbing mole-

Figure 6.10
Artificial UV light can cause valuable works of art to fade. Reproduced from rsc.org, with permission.

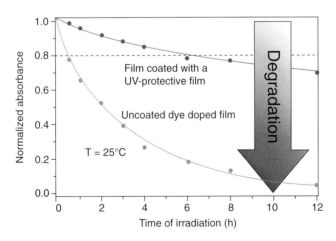

Figure 6.11
Effect of a thin film of organosilica doped with Rhodamine on degradation by UV light. Reproduced from rsc.org, with permission.

cules are mainly responsible for absorption in the UV range as the degradation curve of a fluorescent Rh-101 film coated with an ORMOSIL film with the same composition of the protective coating but *without* the UV absorber molecules is only slightly slower (due to the reflection of light on the surface of the protective coating) than that of the uncoated Rh-101 film. Coatings just 1 μm thick are highly stable upon prolonged exposure to UV light and are fully transparent in the visible region of the spectrum. This means they can be used to coat a wide range of materials, without affecting the way they look.

Similarly, in the crucially important field of photovoltaics (PVs), a single-layer, low-refractive, and cost-efficient antireflective coating based on SiO_2 serves as an optimal alternative to common multilayer compositions employed as cover sheets for photovoltaic modules (Figure 6.12) [10].

When applied on glass using common wet coating techniques (dip-coating, spin-coating, etc.) the resulting coating exhibits up to 50% porosity. Accordingly, the obtained refractive index is between 1.25 and 1.3, which corresponds to a transmission maximum of up to 99% and a solar transmission (weighted average of transmission over total solar range) of >95%; this results in an overall peak power increase of 3.5% of the original performance. Beyond solar applications (glass and hot water collectors), special high-transmittance glass is highly desired for architectural glazing, which requires antireflective glass, and so large companies such as Merck have commercialized several sol-gel silica-based formulations for low-refractive coatings.

Finally, polycondensates with some percentage of Al alkoxides with phenyl- and epoxyfunctionalized alkoxysilanes (at Fraunhofer ISC) [11] are excellent barrier coatings for the coating film for solar cells integrated into modules. Easily deposited by spraying, and following curing at relatively low temperature (70–80 °C), the latter sol-gel coatings provide improved resistance to radiation damage. The light weight, durable sol-gel coating reduces manufacturing steps and the overall weight of traditional Si-based solar cell arrays because the hybrid sol-gel layer replaces the heavy, fragile glass coverings normally applied to the surfaces of the solar cells.

In the former case, the *flexible* nature of the encapsulants results in an optimized encapsulation process that was specially developed for manufacturing of flexible thin film solar cells whose lower cost (compared to traditional Si-based cells) and low material demand is crucial for future dissemination of PV. A low final cost of the solar modules indeed requires the extremely high productivity typical of roll-to-roll manufacturing, whereas flexible modules are suited for non-planar surfaces as well as for wide application in the building industry where such flexible PV cells will soon become a standard integrated part of construction components [12]. The new "one component encapsulant" developed can protect the solar cell from water vapor in the atmosphere, having the necessary very high barrier properties (Figure 6.13). The Samsung SDI AMOLED display (Figure 6.14), 50 μm thick, is currently the world's thinnest (amoled.samsungsdi.com).

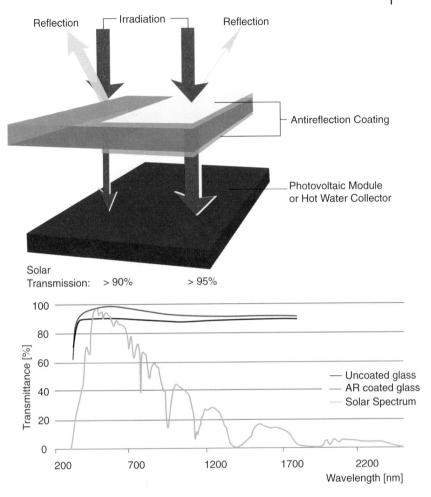

Figure 6.12
The solar transmittance of a glass sheet coated with a sol-gel porous SiO₂ antireflective layer increases from 90% to 95%, resulting in an overall 3.5% increase in the PV conversion efficiency. Image courtesy of merck.com.

The Barix encapsulation and barrier film technology of the US company Vitex Systems (www.vitexsys.com) has enabled the actual production of such a display that, being made of OLED, requires the highest protection against humidity and oxygen penetration. The technology is, once again, based on solution (nano)chemistry and allows the rapid creation of a multilayer barrier (Figure 6.15), affording coatings typically less than 3 μm thick.

This is achieved by coating the substrate surface with a liquid monomer that fills in all the valleys and submerges all the peaks, creating a microscopically flat surface. The liquid is then hardened (polymerized) into a

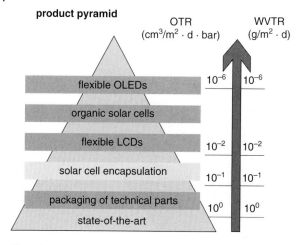

Figure 6.13
Especially for high value added applications like encapsulation of solar modules, the requirements for oxygen and water vapor barrier are orders of magnitude greater than today's state of the art polymer technology. Image courtesy of the Fraunhofer ISC.

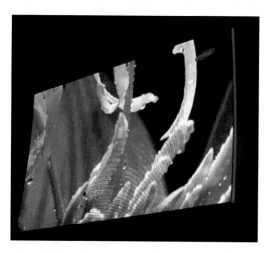

Figure 6.14
The Samsung SDI AMOLED display is currently the world's thinnest. Image courtesy of Samsung SDI.

solid polymer film. A first layer of a transparent ceramic is then deposited to create the first barrier, and a second polymer layer is applied to protect the barrier and create a second flat surface.

This barrier/polymer combination is repeated many times until the *desired level* of water and oxygen impermeability is achieved. Each inorganic layer is a near-perfect barrier film, and the intervening polymer layers de-

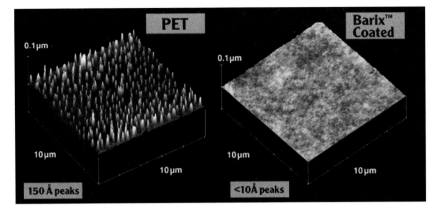

Figure 6.15
The Barix coating combines a smooth surface with redundant ceramic barrier layers and a protective overcoat. Image courtesy of Vitex Systems.

couple what few defects there are, producing a barrier film that meets the requirements of LCD and OLED displays. This composite structure is made up of layers that are so thin that they are flexible enough to be rolled up and still retain their properties.

References

1 Choate, T.F. (1932) Dendritic polyurethane coating, US Patent 4273855.
2 Pagliaro, M. (2009) *Silica-Based Materials for Advanced Chemical Applications*, RSC Publishing, Cambridge.
3 Sanchez, C. and Gomez-Romero, P. (2004) *Functional Hybrid Materials*, Wiley-VCH Verlag GmbH, Weinheim.
4 Detty, M.R., Drake, M.D., Tang, Y., and Bright, F.V. (2007) Hybrid antifouling coating compositions and methods for preventing the fouling of surfaces subjected to a marine environment, US Patent 7,244,295 B2.
5 Mahltig, B., Haufe, H., and Böttcher, H. (2005) *J. Mater. Chem.*, **15**, 4385.
6 Haufe, H., Muschter, K., Siegert, J., and Böttcher, H. (2008) *J. Sol-Gel Sci. Technol.*, **45**, 97.
7 Mahltig, B. and Textor, T. (2008) *Nanosols and Textiles*, World Scientific, Singapore.
8 Bescher, E., Piqué, F., Stulik, D., and Mackenzie, J.D. (2000) *J. Sol-Gel Sci. Technol.*, **19**, 215.
9 Zayat, M., Garcia-Parejo, P., and Levy, D. (2007) Preventing UV-light damage of light sensitive materials using a highly protective UV-absorbing coating. *Chem. Soc. Rev.*, **36**, 1270.
10 Kursawe, M., Anselmann, R., Hilarius, V., and Pfaff, G. (2005) Nanoparticles by wet chemical procession in commercial applications. *J. Sol-Gel Sci. Technol.*, **33**, 71.
11 Houbertz, R., Schulz, J., Froehlich, L., Domann, G., and Popall, M. (2003) *Mater. Res. Soc. Symp. Proc.*, 769.
12 Pagliaro, M., Palmisano, G., and Ciriminna, R. (2008) *Flexible Solar Cells*, Wiley-VCH Verlag GmbH, Weinheim.

7
Better Medicine Through Nanochemistry

7.1
Nanomedicine

Nanochemistry in medicine aims to create materials and devices capable of addressing medical problems, using chemical knowledge to maintain and improve human health at the nanoscale. Beyond diagnosis and treatment (therapy) of medical conditions, modern medicine is also interested in disease prevention [1]. Accordingly, nanomedicine – the medical application of nanotechnology – will have far-reaching implications in each of the three main aspects of the medical profession. However, there is little reason for separating nanomedicine research from other types of medical or nanotechnological research even if, for example, nanomedicines are a new class of pharmaceuticals requiring a new regulatory process: what will "ultimately determine the level of recognition, is the size of the market that nanomedicine applications can potentially reach within established sectors of the industry" [2].

Thus far, most nanotechnology applications to medicine include nanobased *drug delivery* systems, and *diagnostic* tests [3]. From an analytical viewpoint, nanotechnology is important for biology because biological reactions occur at micro- to nanomolar concentrations; consequently, if we want to detect interactions between proteins and DNA or molecules and cells, we need to make very small volumes or very small structures for *in situ*, *in vivo* real time analysis.

Chip technology means we can use very small volumes, making nanotechnology key for this kind of work. Companies such as Immunicon Corporation commercialize chip-based analytical nanotechnologies. Many of these already commercialized products are not available directly to the consumer. Instead, they are used by researchers involved in drug discovery, physicians in need of better imaging techniques, and as prescriptions to treat particular kinds of illness.

The two timelines shown in Figure 7.1 illustrate estimated commercialization timeframes for a select set of nanotechnology drugs, delivery systems,

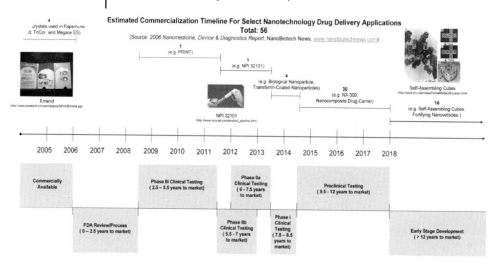

Figure 7.1
Timeframes based on the number of years that FDA regulated products generally take to get through each stage of the development process. (Early stage development, preclinical testing, phase I clinical testing, phase IIa clinical testing, phase IIb clinical testing, phase III clinical testing, and FDA review/process.) Source: nanotechproject.org.

diagnostic tests, and devices that are currently being developed, from applications that are in early stage of development to ones that are already in latter stages of clinical trials. The vast majority of items fall within the early stage development and preclinical testing phases. These applications are just a fraction of the actual number of nanotechnology products currently in the development pipeline; however, this trend is one that is seen for products in most branches of nanomedicine. The manufacturing company and the name of each of these commercialized products are listed below (www.nanotechproject.org):

Appetite control
Megace ES, Par Pharmaceutical Companies, Inc.

Bone replacement
Zirconium Oxide, Altair Nanotechnologies, Inc.

Chemical substitute
Neowater®, Do-Coop Technologies Ltd.

Hormone therapy
Estrasorb, Novavax, Inc.

Immunosuppressant
Rapamune, Wyeth

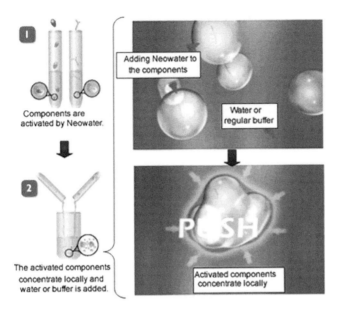

Figure 7.2
How Neowater works. Image courtesy of Do-Coop Inc.

Medical tools
TiMESH, GfE Medizintechnik GmbH

Each of this technology has a large potential to create change. For example, Neowater commercialized by Israeli company Do-Coop Technologies (www.docoop.com) is a stable system of largely hydrated nanoparticles, like non-ionic detergent derived micelles, that can reduce the entropy of aqueous solutions. Unlike traditional nanotechnology, which focuses on a nanoparticle end product, Neowater builds upon the unique ability of nanoparticles to modify the physical properties of water molecules around them (Figure 7.2).

Water kept at 4 °C is first irradiated with a low power radio-frequency signal of about 900 MHz, after which a micron-sized powder of barium titanate ($BaTiO_3$), heated to about 900 °C, is dropped into the water. The RF irradiation continues for an additional 5 min. The water is then held at room temperature for about two days during which time most of the source powder (containing the larger particles) sediments to the bottom and the clear part of the water is collected separately. According to the amount of Ti in the "doped" water and the typical size of the nanoparticles (5–50 nm) its density is nanomolar, that is, below 10^{-12} particles per liter.

Each nanoparticle within Neowater, with its huge surface, creates a "surface effect" and in turn organizes the water molecules surrounding it. This is similar to the surface effect of organelles within living cells. Both organelles and the nanoparticles use this unique mechanism to create intracellular water. While the former is within cells and organs and cannot be harnessed or used in a laboratory bench, the latter can.

As a result, hydration of compounds treated with Neowater changes, resulting in increased stability and bioavailability of hydrophilic and hydrophobic compounds and thus in better drug delivery. Drug development can be enhanced and expedited in an efficient way when utilizing Neowater in the process of compound hydration as it enables superior water-based biocatalysts, solvents, reagents, media enhancers, and buffers that maximize the efficiency of new and existing products and processes. From a biopharma point of view, Neowater can enhance protein stabilization and, among others, enzyme stabilization. Utilizing stem cell media (or standard cell media) based on Neowater can enhance stem cell growth, cell line growth, stem cell cryopreservation, and cell line cryopreservation.

7.2
Hemostasis: Change in Surgery and Emergency Medicine

In 2006 the team of Rutledge Ellis-Behnke, a researcher at Boston's MIT Department of Brain and Cognitive Sciences, created a liquid that looks exactly like water but when applied directly onto injured tissue halts bleeding in any tissue in a matter of seconds [4]–a discovery that has the potential to revolutionize surgery and emergency medicine. Hemostasis is a major problem in surgical procedures and after major trauma. Surgeons currently spend up to 50% of their time during surgery packing wounds to reduce or control bleeding, so if the liquid works it would make a profound difference. There are few effective methods to stop bleeding without causing secondary damage. This novel therapy stops bleeding without the use of pressure, cauterization, vasoconstriction, coagulation, or crosslinked adhesives.

The active agent is a self-assembling synthetic peptide that establishes a nanofiber barrier to achieve complete hemostasis immediately when applied directly to a wound in the brain, spinal cord, femoral artery, liver, or skin of mammals. Under conditions like those inside the body, the peptides self-assemble into a fibrous mesh that to the naked eye appears to be a transparent gel (Figure 7.3).

The self-assembling solution is nontoxic and nonimmunogenic, and the breakdown products are amino acids, which are tissue building blocks that can be used to repair the site of injury. Even more remarkably, the material creates an environment that may accelerate healing of damaged brain and spinal tissue.

The mechanism is a kind of molecular band aid. Once the liquid touches an internal organ, it forms a gel; the amino acids assemble into fibers and stop the bleeding. The degradable peptide then breaks down into nontoxic

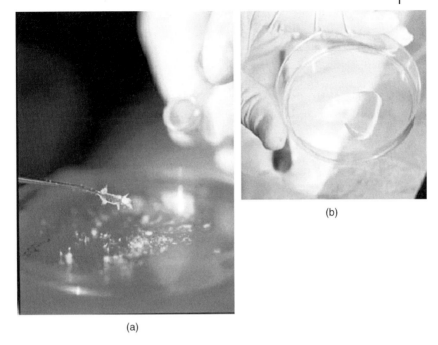

Figure 7.3
(a) Rutledge Ellis-Behnke mixes a powder of short, engineered chains of amino acids with deionized water to form a clear solution. (b) The clear fluid in this dish transforms into a gel in the presence of blood; such a gel can stop bleeding almost instantly. Image courtesy of Asia Kepka, asiakepka.com.

products as the tissue heals. These products can even be used by cells to rebuild damaged tissue. Arch Therapeutics (www.archtherapeutics.com) was founded in mid-2006 to develop the material for commercial use after licensing the technology from MIT. Clinical trials have already begun. Promising results in an animal model mean that human trials could begin in as little as three years. Once costs are low enough, this material will even become part of home first-aid kits.

7.3
Biogels: Biotechnology Made Possible

Enzymes, antibodies and whole cells as well as polysaccharides, phospholipids and nucleic acids and all sorts of small and large organic molecules can be entrapped within the immense porosity of glass (silica, or SiO_2) using the sol-gel a process. The resulting highly porous glass works both as support and as *protective* envelop of the entrapped species. Indeed, practical biotechnology applications require such effective immobilization of biologicals over solid supports. Organic polymers are not stable enough, whereas glass is; and, furthermore, polymers swell in water, making impossible application. By associating a hard and resistant material such as glass with biologicals, the new process has created a new paradigm, merging bio- and materials chemistry. Not only in fact is the biological activity of the entrapped species retained, it also improves activity in comparison to solutions. Hence, not only life can develop within glass but life processes can be made even more efficient and biologicals resilient, to be exploited ultimately to human ends.

A major example of the possibilities opened up by sol-gel nanotechnology concerns gene therapy. In 2005, organosilica nanoparticles – instead of traditionally employed virus – were shown to be excellent candidates for efficient DNA delivery in the brain [5]. ORMOSIL-mediated delivery does not cause the tissue damage or immunological side effects that have been commonly observed with viral-mediated gene delivery. Hence, in place of dangerous viruses traditionally used as the gene delivery vehicle, ORMOSIL amino-functionalized nanoparticles can efficiently introduce a healthy version of the gene into the patient, thus opening up a route to gene therapy that involves treating diseases caused by missing or defective genes, and even to repairing neurological damage caused by disease, trauma, or stroke.

The cationic amino groups at the ORMOSIL's cage surface readily bind with negatively-charged DNA plasmids and the resulting doped material, using plasmids that contained a gene coding for enhanced green fluorescent protein (EGFP), injected into various regions of the mouse brain was even more effective than some viral vectors at delivering the EGFP gene to nerve cells without any damage to the mice nerve cells (Figure 7.4).

Release of DNA *in vivo* takes place due to the increased acidic conditions inside living cells, which results in destabilization of the ORMOSIL–DNA complex. SiO_2-based nanoparticles, in fact, do *not* release encapsulated biomolecules because of the strong hydrogen-bonding between the biomolecule's polar centers and the silanols at the cage surface (as ORMOSIL-entrapped hydrophobic molecules are *not* leached in aqueous system due to strong hydrophobic interaction) [6].

Hence, by using an amino-functionalized ORMOSIL of intermediate HLB (hydrophilic–lipophilic balance) to entrap amphiphilic molecules such as DNA, the plasmids bound to the amino-functionalized nanoparticles are completely protected against enzymatic degradation due to hindered access of the enzymes to the DNA that is immobilized at the cage's surface, while giving a controlled release behavior [7]. As already noted, in the near future, this technology may make it possible to repair neurological damage caused by disease, trauma, or stroke.

Figure 7.4
Organically modified silica nanoparticles: a non-viral vector for *in vivo* gene delivery and expression in the brain. Reproduced from Reference [6], with permission.

7.4
Small is Beautiful? Nanotech Cosmetics

Paris-based cosmetics giant L'Oreal often repeats the claim that in the early 2000s it was the top nanotechnology patent holder in the United States with more than 200 patents: a figure that shows the relevance of nanotechnology to the $60 billion international beauty products industry. The beauty industry is adding "nanoparticles" to lipstick, foundation, and anti-aging products because at the nanoscale familiar substances have novel optical and biological properties. For example, as larger particles titanium dioxide and zinc oxide are white and opaque. But at the nanoscale, the same substances become transparent, enabling their use in moisturizers and foundations [8].

L'Oreal's first nanotechnology product was introduced in 1998: the anti-wrinkle cream labeled Lancome Resurface using polymer nanocapsules to incorporate vitamin A and deliver the active ingredients deeper into the skin. The company has thus patented the use of dozens of different "nanosome" particles, and today commercializes creams that deliver vitamin C into skin (Revitalift) and other facial moisturizers (Vichy Reti C and Biotherm Age Fitness Nuit) (Figure 7.5).

Figure 7.5
The whole product line of anti wrinkle creams Revitalift of L'Oreal makes use nanosome particles entrapping retinol. Image courtesy of L'Oreal.

"Nanosomes" or nanoscale liposomes are small droplets of liquid enclosed in a nano-thick shell. They are essentially a delivery mechanism used in cosmetics to deliver active chemicals deeper into the skin and in some nutrient supplements for enhanced absorption. The capsules act like sponges that soak up and hold the product inside until the outer shell dissolves. The 200 nm particles penetrate the surface of the skin through the pores where they dissolve and provide benefit.

Dior has invented the "liposome" to perform the same function; and Shiseido commercializes a silica powder entrapping enzymes (Elixir Skinup) in which the enzyme is slowly released to provide relief from dry, rough skin. Other large brands like Revlon, Avon, Prestige, and The Body Shop also sell nano-cosmetics.

Conventional skincare products form a barrier to prevent moisture loss. The miniaturized particles, by contrast, are absorbed *deeper* into the skin than more traditional treatments and boost production of new cells so that the skin remains soft and free of wrinkles.

With millions of consumers using cosmetic products as part of their daily routine, one would expect that products were thoroughly assessed for safety. Unfortunately, the health effects of nano-cosmetics remain poorly understood and effectively unregulated [9]. Although potentially hundreds of products are on sale right now, companies are still not required to label nano-ingredients [10]. The increased capacity of nanoparticles to penetrate skin and gain access to our bodies' cells may be useful for medical purposes, but it could also result in greater uptake of substances that have a negative health effect. The cosmetics industry argues that risks for consumers are low, as there is no evidence that nanoparticles in cosmetics penetrate healthy, intact adult skin.

Yet nanotechnology can go further and actually provide products that are safer and healthier than traditional products. In general, in fact, nanoparticle usage broadens the range of chemicals that can be used in cosmetics and in medicine. They are used to coat the surfaces of microscopic packages of vitamins, growth promoters, and other substances that, if used in their raw form, would cause *irritation*. When modified, they can be taken into the underlying layers of the skin.

For example, Sol-Gel Technologies (www.sol-gel.com) in Israel manufactures microcapsules made of sol-gel entrapped benzoyl peroxide (BPO) for effective, non-irritating treatment of acne. BPO crystals are encapsulated in transparent, porous silica shells by carrying out the sol-gel hydrolytic polycondensation in an emulsion phase (Figure 7.6).

The inert core shells serve as a safe protective barrier, preventing direct contact between the BPO and the skin and significantly reducing side effects. The product (Cool Pearls BPO) then precisely controls the amount of active ingredient deposited over time, adjusting it to meet the skin's individual needs.

The mechanism involves migration of the skin's natural oily secretions through the silica pores into the capsule. The oils dissolve the BPO crystals and carry the BPO to the sebaceous follicles. The amount of skin lipids

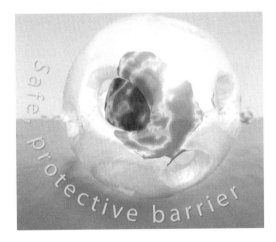

Figure 7.6
Anti-acne drug BPO crystals once entrapped in a silica microcapsule shells exert the medical function without irritating the skin. The product Cool Pearls™ Anti Acne Kits was commercialized in 2009. Image courtesy of Sol-Gel Technologies.

controls the rate at which the BPO is released, delivering optimal benefits. Applied to the skin in simple formulations, in fact, BPO also produces numerous side effects, ranging from skin irritation, stinging, itching, peeling, and dryness to redness. The acne therapy market is worth $1 billion and the company in 2008 entered into a development and licensing agreement with a US pharmaceutical company for which it will receive royalties from sales in the US and an extra $24.7 million to fund further product development aimed at other dermatology market segments in which BPO is widely employed.

Similarly, SGT manufactures inert glassy silica micro-particles encasing high concentrations of sunscreen molecules within a thin shell of inert sol-gel glass. With increasing public knowledge that ultraviolet (UV) sunlight is the primary cause of skin aging, wrinkles, and skin cancer, people use sunscreens more often, in higher concentrations (high SPFs), and in daily wear cosmetic preparations. A direct consequence of the increasing use of sunscreen molecules is that an increased amount of these molecules may penetrate through the epidermis into the body. Moreover, when UV light is absorbed by the sunscreen molecules, photodegradation products, including free radicals, may be formed and interact with body tissues. As the UV filters are encapsulated in glass micro-particles (Figure 7.7a), a protective and homogeneous UV-absorbing layer is placed on the skin's surface. The glass walls prevent interaction between the UV filters and skin. In other words, the encapsulated UV filters predominantly remain on the surface of the skin. Consequently, Eusolex® UV-Pearls™ (Figure 7.7b) are now used in the top formulations of the ever-increasing sunscreen market segments for sensitive skin, both young and aged.

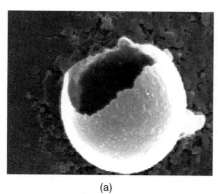

(a)

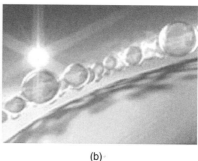

(b)

Figure 7.7
(a) A broken, spherical silica particle entrapping an Active Pharmaceutical Ingredient (API) has 85% free volume. (b) Such particles are used in formulations such as Eusolex UV-Pearls that reduce dermal uptake compared to free UV filters – thus they do not irritate the skin, affording new application possibilities for hydrophobic UV filters. Photograph courtesy of Sol-Gel Technologies Ltd.

The sol-gel microencapsulation here enables a high load of ingredient (e.g., 85% of the weight of the particles) to be incorporated into a preparation and thus achieve sustained delivery of the active ingredients to the skin under defined mechanical or chemical conditions. In addition, it affords low leaching (i.e., non-delivery) and enables the isolation of a component from its surrounding ingredients and the use of incompatible ingredients in the same formulation, because mineral and amorphous coating silica offers better tightness and resistance to extraction forces than polymers or waxes.

The formulation of micro-capsules trademarked Eusolex UV-Pearls has been commercialized by Merck. The products are supplied as aqueous dispersions containing approximately 35% in weight of the UV absorber. The white liquids contain Eusolex UV-Pearls of about 1.0 µm diameter on average; 90% of the capsules are <2.5 µm in diameter; and are thus sufficiently small to be transparent when applied to the skin and to give a pleasant feeling. The aqueous microemulsion dispersion provides new opportunities for cosmetic formulators as oil-soluble organic sunscreens can now be incorporated into the aqueous phase, and incompatibilities between cosmetic ingredients can be prevented to the benefit of novel combinations in *one* cosmetic product.

7.5
Nanotechnology in Orthopedics

Several biomaterials in the form of self-assembled nanofibers/nanoparticles, nanofibers, and nanocomposites are already being used as integral parts of biomedical devices to improve their *in vivo* performance (Figure 7.8) [11]. For instance, some of the nanoparticle composites available for dental filling and repair applications include Filtek Supreme, Ceram X duo, Tetric EvoCeram, Premise and Mondial.

Filtek Supreme, for example, is a restorative nanofiller composite material of the US company 3M, combining the esthetics required for anterior restorations with the strength needed for posterior restorations (Figure 7.9).

At the nanoscale, materials possess several novel properties. For example, SEM pictures of the Filtek composite compared to those of other commercial microfillers (Figure 7.10) show how the clusters of nanometer-sized particles provide the strength and wear properties of a hybrid while maintaining the polishability and polish retention of a microfill.

In comparison, other composites are obtained by grinding down large particles to create smaller ones. This results in an unpredictable particle size, with the larger particles leading to undesirable esthetic properties.

Several other nanohydroxyapatite (nHAp)-based products such as Ostim, Vitoss, and Perossal are now commercially available for bone filling,

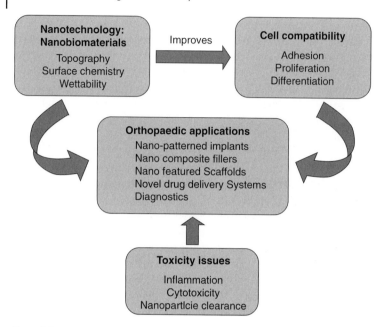

Figure 7.8

Nanotechnology and orthopedic applications. Reproduced from Reference [11], with permission.

whereas other nanotechnology approaches being explored include coating of orthopedic implants with nHAp and bioactive signaling molecules to stimulate osteoblast proliferation and differentiation. Bone in fact is a nanocomposite material that consists of hierarchically arranged collagen fibrils, hydroxyapatite, and proteoglycans on the nanometer scale. Cells thus are accustomed to interacting with nanostructures, and the nano chemistry bottom-up approach clearly has great potential to create natural bone-like environments that potentially enhance bone tissue regeneration/repair.

The overall aim is to fabricate nanostructures that simulate the native hierarchical structure of the bone, showing sufficient strength and flexibility as well as being biocompatible, which requires not only lack of toxicity but also a suitable structure to enable cell proliferation and efficient removal of nutrients and metabolic wastes. The latter properties are being addressed through the fabrication of nanofibers, namely with the advent of nanofabrication techniques. In other words, nanochemistry is opening the route to novel biomaterials that will soon complement the already abundant offer of nanomaterials for better orthopedics. Thorough evaluations of these structures in suitable animal models these biomaterials are on course to make the transition to clinical use.

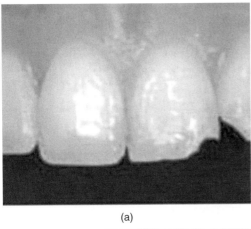

(a)

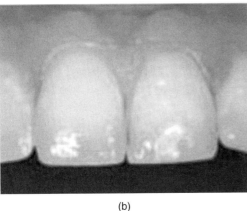

(b)

Figure 7.9
A trauma case (a) restored with Filtek™ Supreme Plus restorative nanofiller (b); the results are evident from these clinical photographs Photograph courtesy of Dr. Manhart, University of Munich, Germany.

7.6
A Hybrid, Welcome Science

Nanomedicine is a highly cross-disciplinary field that requires the integration of disciplines such as biology, medicine, materials science, physics, and manufacturing. Accordingly, perspectives must broaden and avoid a monodisciplinary focus. The new journal *Wiley Interdisciplinary Reviews: Nanomedicine and Nanobiotechnology* provides exactly the type of perspective that a professional in an adjacent area of research needs to help both find his/her place in the field and remain updated in a rapidly evolving field in which is difficult to keep current. In addition, the highly structured format

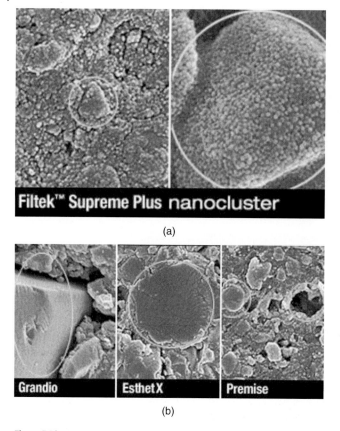

Figure 7.10
In Filtek Supreme Plus the nanocluster consist of individual nanoparticles 20–75 nm (a); the larger nanoclusters do not pluck out from the restoration but the individual nanoparticles do, leading to better long-term polish retention; in the case of other commercial fillers (b), larger particles increase surface roughness and negatively affect long-term polish retention. Photograph courtesy of 3M.

of the publication allows instructors to create class texts for different disciplines (such as medicine or bioengineering or materials science), providing the novice with a means to enter the field and gain familiarity with many aspects that would have previously required diverse compilations of single articles from different sources.

Indeed, we agree with Berube that nanomedicine products will actually help anchor public sentiment towards nanotechnology in general in a very positive fashion [12], as they will be perceived as *archetypal* for nanotechnology based on public expectations of technological development in the field of medicine in general. People everywhere welcome the introduction of any new technology that serves to improve health and prolong life expectations:

Figure 7.11
The spray for sealing glass and ceramic surfaces Magic Nano was recalled by the distributor after a public warning issued by the manufacturer. Photograph courtesy of europa.eu.

"Sating this apprehension makes the adoption of this technology, including nanomedicine, more likely against the same technology associated with some less essential application, like stronger automobile bumpers and better performing baseball bats, as most of current nanotechnology product releases have involved the use of nanoparticles in coatings (paint), and as reinforcement when associated with another media (composite materials)" [13].

On March 27, 2006 a bathroom cleaner called Magic Nano went on sale in Germany. Three days later it was withdrawn from the market after nearly 80 people reported severe respiratory problems and six were admitted to hospital with fluid on their lungs [13] (Figure 7.11).

Although most of the symptoms soon cleared up, critics of nanotechnology have been quick to identify this as one of the first examples of a sinister technology. Kleinman's Magic Nano was thus the first nano-product to be recalled from the market, even if the product was later shown to contain no nanoparticles [14]. At about the same time, marketing of Samsung's SilverCare washing machine (Figure 7.12) product line, which releases silver ions (Ag^+) nanoparticles to disinfect (sanitize) clothing without the need for either hot water or bleach, was curtailed for a while in the USA after the EPA indicated it intended to regulate silver-ion generators as pesticides.[1]

1) The EPA officially announced its position on silver ion generators in the September 21, 2007 Federal Register.

Figure 7.12
Samsung's SilverCare washing machine was curtailed for a while last year in the USA after the EPA indicated it intended to regulate silver-ion generators. Photograph courtesy of Samsung.

However, both of these examples did not receive substantial media attention, hence they were not amplified; in addition, patient consumers will not be reluctant to embrace nanomedicine because effective nanomedicine products are already helping patients in several ways, such as those mentioned above.

In general, careful development, testing, and monitoring of new medical products that contain nanotechnology will of course be essential in developing and maintaining public as well as investor confidence. Liposome-encapsulated drugs produced by Nexstar (doxorubicin for cancer treatment and amphotericin B for fungal infection) had sales over $20 million already in 1999 [15]; yet, it will be gene therapy employing sophisticated organosilica nanostructures as synthetic enclosures for DNA and virus delivery delivering genetic material into target cells that will make nanomedicine products a new, huge commercial success.

It can be envisaged that the production of nanomaterials will make use of advanced fabrication methods that are analogous to actual fabrication methodologies, affording control and flexibility in the engineering of designed particles such as PRINT (particle replication in non-wetting

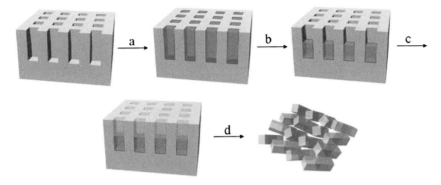

Figure 7.13
Fabrication of Janus particles using PRINT technology: (a) one monomer solution that has been diluted by a volatile solvent is filled into the mold; (b) after evaporation of the solvent, the remaining liquid is partially photo-cured; (c) the other monomer solution is filled, all liquids are completely cured and (d) Janus particles are obtained after harvesting and purification. Reproduced from: Zhang, H., Nunes, J. K., Gratton, S. E. A., Herlihy, K. P., Pohlhaus, P. D., DeSimone, J. M. (2009) Fabrication of multiphasic controlled size and shape. New J. Phys., 11, 075018.

templates, Figure 7.13), for producing precisely controlled polymeric vectors in the tens of nanometers to the micrometer size range, well beyond the possibilities offered by liquid-phase chemistry techniques employed to prepare traditional carriers such as liposomes, dendrimers, and colloidal particles [16].

The general commercialization dynamics of medically oriented nanotechnology (Figure 7.14) differs from that of other nanotechnologies due to the need for a continuous focus on efficacy and safety that requires nonclinical and clinical studies as well as demonstration of the ability to safely manufacture the product.

Funding is the most critical aspect, and it is perhaps not surprising in light of the nanotech hype that dominated the late 1990s and the early 2000s that, today, there is more discussion going on rather than doing.

Thus far, funding of successful nanomedical companies has either come from governments or from private equity investors financing the creation of small companies licensing technology from research institutes for commercialization and development. Private investment has been particularly low in the United States as well as in Europe, with small companies emerging in more peripheral markets such as in Israel and Canada. Investments in this sector will greatly benefit from the entrance of new financial global players such as the Islamic finance institutions due to the large beneficial impact of medical products (which is a precise requirement guiding the

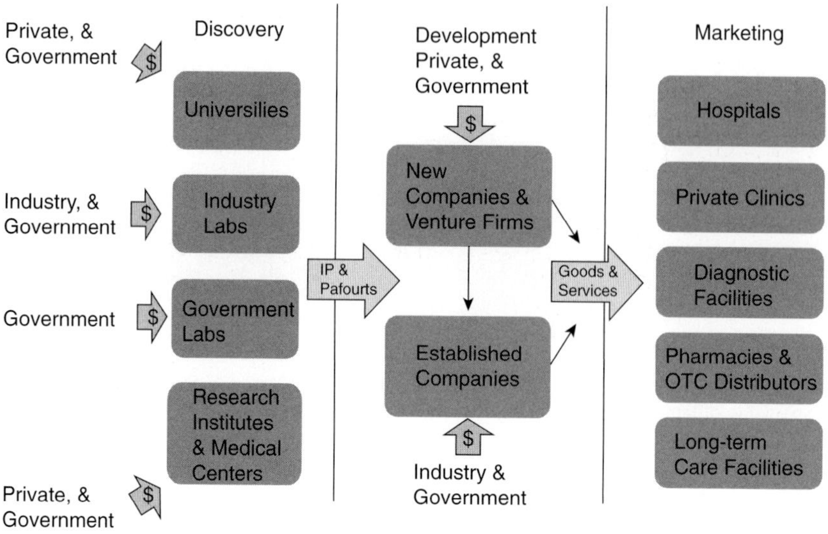

Figure 7.14
Commercialization discovery, development, funding, and distribution dynamics for medical nanotechnology. Reproduced from Reference [15], with permission.

investing options for these institutions) as well as to the historic crisis in the global financial system in 2008 that saw the bankruptcy of Lehman Brothers and the public bailout of some of the world's largest banks both in Europe and in the USA.

References

1 Freitas, R.A. Jr. (2005) What is nanomedicine? *Nanomed. Nanotech. Biol. Med.*, **1** (1), 2–9.
2 Editorial (2006) Nanomedicine: a matter of rhetoric? *Nat. Mater.*, **5** (4), 243.
3 National Health Information, LLC (2006) Nanomedicine, device & diagnostics report. NanoBiotech News data.
4 Ellis-Behnke, R.G., et al. (2006) Nano hemostat solution: immediate hemostasis at the nano scale. *J. Nanomed.*, **2**, 207.
5 Bharali, D.J., Klejbor, I., Stachowiak, E.K., Dutta, P., Roy, I., Kaur, N., Bergey, E.J., Prasad, P.N., and Stachowiak, M.K. (2005) Organically modified silica nanoparticles: a nonviral vector for *in vivo* gene delivery and expression in the brain. *Proc. Natl. Acad. Sci. U.S.A.*, **102**, 11539.
6 Roy, I., Ohulchanskyy, T.Y., Bharali, D.J., Pudavar, H.E., Mistretta, R.A., Kaur, N., and Prasad, P.N. (2005) Optical tracking of organically modified silica nanoparticles as DNA carriers: a nonviral, nanomedicine approach for gene delivery. *Proc. Natl. Acad. Sci. U.S.A.*, **102**, 279.
7 Luo, D. and Saltzman, W.M. (2006) Nonviral gene delivery: thinking of silica. *Gene Ther.*, **13**, 585.
8 Lewis, W. (2004) *Beauty Secrets: An Insider's Guide to the Latest Skin, Hair and Body Treatments*, Quadrille Publishing, London.

9 Rogers, L. (2005) Safety fears over "nano" anti-ageing cosmetics. The Sunday Times (July 17).
10 Miller, G. (2009) Nano cosmetics may carry big risks. The Age (March 28).
11 Laurencin, C.T., Kumbar, S.G., and Prasad Nukavarapu, S. (2009) Nanotechnology and orthopedics: a personal perspective. *Wiley Interdiscip. Rev. Nanomed. Nanobiotechnol.*, **1**, 6.
12 Berube, D.M. (2009) The public acceptance of nanomedicine: a personal perspective. *Wiley Interdiscip. Rev. Nanomed. Nanobiotechnol.*, **1**, 2.
13 (2006) Health worries over nanotechnology. The Economist (12 April).
14 von Bubnoff, A. (2006) Study shows no nano in Magic Nano, the German product recalled for causing breathing problems. Small Times (May 26).
15 Hobson, D.W. (2009) Commercialization of nanotechnology. *Wiley Interdiscip. Rev. Nanomed. Nanobiotechnol.*, **1**, 189.
16 Canelas, D.A., Herlihy, K.P., and DeSimone, J.M. (2009) Top-down particle fabrication: control of size and shape for diagnostic imaging and drug delivery. *Wiley Interdiscip. Rev. Nanomed. Nanobiotechnol.*, **1**, 391.

8
Getting There Cleanly

8.1
Why Sustainable Nanotechnology?

According to Lux Research (www.luxresearchinc.com), a consultancy, in 2008 there were about 300 companies working on nanomaterials for wind power, photovoltaics, packaging materials, batteries for efficient and compact energy storage, and a multitude of other products or components. Only five years before that the market was practically nonexistent.

However, nano-products in general have so far escaped public scrutiny, falling through loopholes in government regulation in most developed countries except Canada. Until the early 2000s, nanotechnology's meteoric rise was achieved with little regard to their potential effect on human health and the environment. Subsequent research, however, is clearly showing that nanoparticles are often harmful and toxic. Researchers have discovered that silver nanoparticles used in socks to reduce foot odor are being released in the wash with possible negative consequences. Silver nanoparticles, which are bacteriostatic, may then destroy beneficial bacteria that are important for breaking down organic matter in waste treatment plants or farms [1].

Other studies found that when rats breathed in nanoparticles the particles settled in the brain and lungs, which led to significant increases in biomarkers for inflammation and stress response [2]; some forms of carbon nanotubes if inhaled in sufficient quantities could be as harmful as asbestos [3]. Definitely, we need to adopt a preventive approach with nanomaterials, classifying them first and then using them according to safety procedures.

Canada is the first country in the world to have regulated nanomaterials. In a pioneering effort in this country, at the frontier in nanotechnology, companies and institutions that manufactured or imported more than 1 kg of a nanomaterial in 2008 will be required to submit all of the information they have – physical and chemical properties, toxicological data, and methods of manufacture and use. The government agencies, Health Canada and

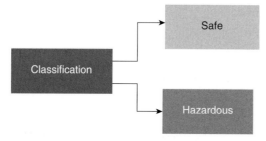

Figure 8.1
Classification of nanomaterials started in Canada in 2009.

Environment Canada, will use the information for the assessment and future risk management of nanomaterials. Slowly, governments in Europe and in the USA are eventually financing the research of scientists who are trying to establish predictable paradigms of toxicity that can help to classify these materials (Figure 8.1).

In the rest of the world, regulators continue to dither over the issue, in much the same way that financial regulators did over private banks for almost two decades until the great crisis of 2008. Self-imposed regulation is thus invoked by "responsible companies" as if "responsibility," and not profitability, were the primary goal of the enterprise. This has arisen even though public bodies such as the Royal Society in the United Kingdom had recommended already in 2004 that, given their risks, all products containing nano-ingredients should pass rigorous safety testing, and face mandatory labeling, before they can be sold; and despite private institutions such as reinsurance agent Swiss Re recommending that "the precautionary principle should be applied whatever the difficulties."

We definitely need to learn lessons from past beauty and chemical product disasters to ensure that we avoid repeating them. Hence, the environmental impact of nanomaterials must be assessed, from many different points of view: fate, transport, waste, toxicity, biomagnification in the food chain. There are no specific regulations on nanoparticles except existing regulations covering the same material in bulk form. Difficulties abound in devising such regulations, because of the likelihood of different properties exhibited by any one type of nanoparticle, which are tunable by changing their size, shape, and surface characteristics [4].

A green chemistry approach needs to be incorporated into nanotechnologies at source (Figure 8.2). In brief, the potential health effects of nanoparticles, along with medical applications of nanoparticles including imaging, drug delivery, disinfection, and tissue repair, must be assessed *before* the widespread adoption of specific nanotechnologies [5], which calls for specific biological and toxicological testing of nano-sized structures as the toxicity of nanoparticles cannot be assessed by simply testing the material

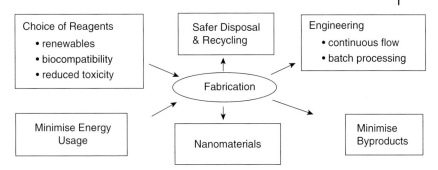

Figure 8.2
Green chemistry considerations in preparing nanomaterials. Reproduced from Reference [5], with permission.

in bulk form. Synthetic nanoparticles inevitably interact with cell membranes, passing through membranes designed to act as barriers, and therefore create many opportunities for toxic effects to occur [6].

Such interactions must be understood and managed so that the many beneficial aspects of nanotechnology give rise to a minimum number of unintended negative consequences [5]. As Cathrine Murphy puts it:

> "The nanotechnology revolution will be an opportunity for a paradigm shift only if 'green' nanotechnology will be embraced by companies designing their new products, with the environment and sustainability in mind. Interestingly, the precepts of 'green chemistry' have been spreading since the mid-1990's, concomitant with advances in nanomaterial synthesis. Eventually, nanomaterial synthesis groups are developing greener, more sustainable production methods, while nanoparticle application groups are exploring sustainable energy sources and environmental remediation as end goals."

> "Overall, we can expect the ideas of sustainability to become part of the design criteria for nanomaterial production and application – especially if the next generation of scientists is already being trained with these ideas in mind" [7].

8.2
Regulating Nanomaterials

A central fact of nanotechnology is that (nano)material properties change with size, not only with composition. Hence, new tools and approaches are necessary to study the scale effects. Moreover, once in the environment, these substances do not act alone. Preliminary data from toxicity tests show that behavior changes with agglomeration, as coatings degrade. As they end up in water, their behavior is going to be much more complex.

The European Commission (EC), the US Environmental Protection Agency (EPA), and the UK Department for Environment, Farming and Rural Affairs (Defra) have all launched their own schemes to gather information on nanomaterials. However, none of them made Canadian-style demands for information from industry, and opted instead for *voluntary* schemes, asking manufacturers to take part and provide them with information about what materials they make. Voluntary schemes, understandably, are not working, with only four companies having so far agreed to participate in the US voluntary scheme and a few more in the UK [8]. The European Commission funds, for example, a "nano observatory" (www.observatorynano.eu).

The Canadian government has opted to set the lower limit for its safety reporting scheme at 1 kg. In Europe, the REACH (Registration, Evaluation, authorization and Restriction of Chemicals) regulations applying to nanoparticles set a regulatory threshold limit at one tonne per year so that if someone is producing 999 kg, a huge number of nanoparticles, that regulation will not cover it; according to a spokeswoman for the environment at the EC, there are no plans to change the regulatory threshold at least until 2012, when REACH will be full reviewed and at which time all recommendations will be considered.

Products modified with nanomaterials have been on the market for almost a decade now. But only in 2008 did researchers from the Swiss Federal Institute of Aquatic Science and Technology report what may be the first detection of an engineered nanoparticle in the environment, an endeavor akin to looking for the proverbial needle in a haystack. The purported source: building facades coated with paints containing nano-TiO_2 (Figure 8.3) [9].

Nano-TiO_2 inhibits the growth of algae and other microorganisms, replacing organic biocides that keep building surfaces clean. In addition, TiO_2 photochemistry also breaks down particulates, which means that nano-TiO_2-coated windows could eventually clean a city's air. Even though fewer biocides are used, the nanoparticles washing off these windows and other coated surfaces into storm drains, streams, and rivers might pose problems for fish and other organisms.

In general, high-throughput testing to enable rapid screening and predictive models of toxicological interactions of nanomaterials need to be developed [10]. Indeed, the variations and dispersities of nanomaterial preparations, as well as the explosive growth in the diversity of the nanomaterials prepared, make conventional toxicological testing limited in value, based on our inability to characterize comprehensively the materials being tested [11]. High-throughput methods have three advantages: (i)

Figure 8.3
Self-cleaning glass roof based on TiO$_2$-nanoparticles. Photograph courtesy of Pilkington Deutschland AG.

speed and efficiency, (ii) direction for and selection of further toxicological studies, and (iii) guidance for syntheses, preparations, and assembly methods. While further detailed studies will nonetheless be required, these can be targeted and focused at key materials and doses, so that more information can be elucidated, including specific biological/toxicological interactions. Likewise, with early feedback, preparation methods can be altered so as to eliminate potentially harmful products and side products.

8.3
Greening Nanomaterials

Green nanotechnology consists of two parts: processes and products. Materials chemists need first to develop processes to make nanomaterials in sustainable ways. Among the benefits, nanoscale manufacturing can reduce the amount of source material that is necessary, get rid of toxic solvents, use less water; and use a reaction temperature close to room temperature. As Wiesner says, concerns over nanomaterials' possible effects on health and the environment have perhaps overshadowed the pressing need to ensure that their production is clean and environmentally benign. Indeed, many of the ingredients used to make nanomaterials are currently known to present risks to human health.

Rather than waiting for regulators, then, the winning nanotech companies will develop products that instead of being harmful will be *beneficial* to the environment, serving either the purpose of producing clean energy or environmental remediation of polluted areas. Providing clean, green energy and safe drinking water for human use, the clean production of metal nanoparticles, and the passivation of nanostructured surfaces are all excellent examples of sustainable nanoscience in action.

8.3.1
Cleaning Up Water

In the quest to clean water of unwanted pollutants, one of the latest tools is shaped like the roots of a tree and can reach 100 nm from tip to tip [12]. This multi-branched molecule is based on a dendrimer, a snowflake-shaped molecule with functionalized junctions that bind targeted contaminants. The team of Mamadou Diallo at the California Institute of Technology has recently produced modified dendrimers that can bind fluoride, chloride, nitrate, bromide, phosphate, and toxic perchlorate. Dendritech Technologies (www.dendritech.com), based in Michigan, is one of only a handful of facilities in the world able to produce dendrimers in large quantities. These nanomaterials might make it possible to tackle seemingly intractable contaminants, such as PCBs. The materials can be recyclable, tailored to specific purposes, and are relatively cheap and are easy to make.

Diallo's team's uneven "dendigraphs" still have a hyperbranched, macromolecular structure and they have a huge binding capacity. In addition, whereas pure dendrimers are too expensive, costing $1000 or more per pound, Diallo and his coworkers can make their less perfect structures for $0.5–15 per pound (Figure 8.4). The resulting water-soluble dendrimers can be used for water recycling, for example, in a "polishing" step designed to be incorporated into existing treatment systems to remove perchlorate. The dendrimers could find their way into household water filters, or be loaded into membranes for groundwater treatment, particularly for concentrating perchlorate.

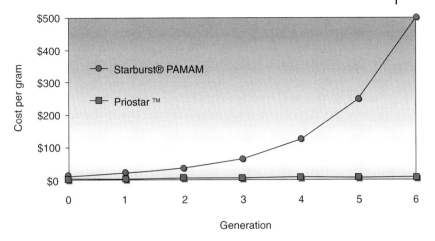

Figure 8.4
Cost profile of Starburst® versus another commercial set of dendrimers. Reproduced from www.dendritech.com, with permission.

8.3.2
Biocompatible Coatings

In the event that a nanomaterial will be exposed to the open environment or to a living organism, the chemical nature of its inner core is not likely to be as important as the chemical nature of its surface [13]. In the context of sustainability and safety, covering potentially toxic surface groups with more innocuous ones appears to be a good strategy; this has the added bonus of improving the biocompatibility of nanomaterials that might be intended ultimately for an *in vivo* application. There are several common strategies one might employ to cover nanomaterial surfaces. Depending on the nature of the nanomaterial, direct chemical covalent derivatization might be performed, retaining the desired physicochemical properties of the inner cores while reducing, for example, cellular uptake of nanorods.

8.3.3
Green Metal Nanoparticles

Gold nanoparticles are of great interest for their optical properties, as well as for unusual and excellent catalytic properties. University of Oregon chemist James Hutchison has developed a greener method to make ultrasmall (1.5 nm) gold nanoparticles, reducing the cost of gold nanoparticles from $300 000 per gram of product to $500 [14]. Hutchison has managed to replace the original reducing agent, a gas, with sodium borohydride, a solid; and the original solvent benzene with toluene. The final purification nanoparticle step, involving liters of organic solvent, can now be performed with a specially engineered filtration membrane.

For the synthesis of metallic nanoparticles, the precursor metal salt in water can be reduced with biomass-derived reducing agents – sugars, glutathione, plant and algal extracts, including starch, extracts from lemongrass – with good control over particle size and shape, and with obvious sustainability advantage (e.g., recycling food waste for chemical production) [15].

8.4
Understand the Risks and Minimize Them

In the global race for nanotechnology leadership, the winners will be those who understand the risks and support the research necessary to minimize them [16]. As put by Thomas, risk assessment of nanomaterials requires adequate characterization of the toxicity potential of nanomaterials and the *exposures* that the population may experience, thereby assessing the exposure potential in occupational settings and for the general population throughout the *life cycle* of nanomaterials (Figure 8.5) [17].

What are the relevant attributes of nanomaterials through which to characterize nanomaterials? They will certainly include particle chemical nature, size, number, and surface area. Yet, an entire metrology remains to be developed. Second, life cycle analysis of nanomaterials in consumer goods and their transformation and degradation in products throughout the life of material is necessary.

The wide range of over 800 diverse consumer products (produced by 484 companies, located in 24 countries), including health supplements, cosmet-

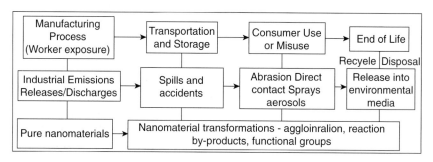

Figure 8.5
Nanomaterial exposure potential through a typical life cycle. Reproduced from Reference [17], with permission.

ics, sunscreens, clothing, electronics, and spray cleaners, in the Woodrow Wilson Center project on emerging nanotechnologies database [18] is clearly indicative of the great range of exposure potential: from very high for cosmetics applied directly to the skin, through to very low for nanomaterials incorporated into durable matrices for electronic devices. In general, we need to learn where nano-products end up, the form in which they exist at the time of exposure and how exposure occurs and what effects result from exposure. In general, the most dangerous nanomaterials would be those that are both *mobile* and toxic [19]. The fullerenes that have been the focus of early toxicity studies are among the least mobile of nanomaterials. Work on nanomaterial mobility in formations resembling groundwater aquifers or sand filters has shown that while one type of nanomaterial may be very mobile a second may stay put. Thus, each nanomaterial behaves differently.

The cycle of fundamental discovery, technological development, revelation of undesirable consequences, and public aversion identified by Wiesner (wiesner.cee.duke.edu) appears unbreakable (Figure 8.6). But it is *not*. We just need to practice prevention and establish a quality management program, based on which we shall explore in advance – before large-scale commercialization of nanomaterials will take place – how environmental risks associated with the production, use, and disposal of these materials can be best managed. Indeed, as environmental engineers and scientists, we have a special obligation to not only look for matches between nanotechnologies and environmental needs but also to anticipate unintended, perhaps negative, consequences associated with the growth of an emerging nanotechnology industry. A critical element of this charge is to perform forward-looking research that explores the possible environmental implications of the products of nanochemistry and their fabrication.

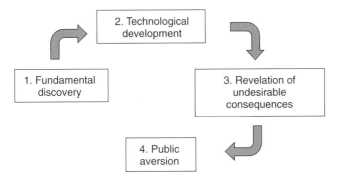

Figure 8.6
"Even disinfecting water – the single most important technological advance ever in prolonging human life – turns out to produce carcinogenic by-products," underlines Mark Wiesner. But the cycle leading to public aversion can be broken!

In practice, and in conclusion, this means that funding agencies and manufacturers of nanomaterials will need to *allocate resources* for real risk assessment studies conducted in tandem with exposure assessments. For example, in the USA a study aimed at investigating the release into the environment of silver from the use of socks containing bactericidal silver nanoparticles clearly indicated that the fabrics release most of the nanoparticles into the environment via solid waste from water treatment plants [20]. Eventually, real studies will tell us which nanomaterials need to be regulated, by discovering which materials are *both* toxic and have a high potential for exposure. In addition, this will eventually lead to the creation of a system of nomenclature for nanomaterials along with the technologies (methods) for measuring materials in practical terms [21].

8.5
Communicating the Nanotech Risk

Public education and awareness programs to inform and educate the business sector, universities, the media, and others on the implications and possibilities that will arise from nanotechnology are simply a necessity [22].

Insights to be used to form strategies for communicating scientifically sound information about nanotechnology in forms that make it accessible to citizens of diverse cultural outlooks is crucial for the very future of the nanotechnology enterprise. Companies engaged in commercializing new nanotechnologies must be aware that media hype has already made nanotechnology a theme of public controversy.

Therefore, companies need to effectively manage the communication process by practicing a careful discipline with their releases of information to the public: not to overpromise the emerging technology and avoid any sort of information release that may have shock value to the media or public.

On the other hand, the *inability* to communicate effectively the potential risks associated with nanotechnology will let nanotechnology suffer the same fate as other disliked technologies, including nuclear power and genetically modified organisms, whose development was stifled by political contention.

In short, if permitted to widen, the disparity between our scientific knowledge of the risks of nanotechnology and our scientific knowledge of *how to communicate* what we know could itself threaten advancement of this important new technology. For example, in a recent experiment conducted jointly by the Cultural Cognition Project (CCP) at Yale Law School and the Project on Emerging Nanotechnologies at the Woodrow Wilson International Center for Scholars (Figure 8.7), a message framing emphasizing the commercial use of nanotechnology to clean up the environment *failed* to

Figure 8.7
Cover of the research brief of a project on emerging nanotechnologies at Yale University. Reproduced from Reference [18], with permission.

produce that effect. Indeed, risk-mitigating framings had the perverse effect of *increasing* nanotechnology risk perceptions overall.

In a climate of cultural conflict, such as that pervading the public discourse in both Europe and the USA, members of the public are much *less* likely to converge on scientifically sound information that benefits society as a whole. Indeed, in such a tendentious climate, those who have a stake in misleading the public (e.g., rallying support in favor of blocking new nanotechnologies that could harm *their* business) can much more readily do so. In a so-called "cultural cognition" attitude people tend to form beliefs about the risks and benefits of an activity that fit *their* cultural evaluations of it. One of the central teachings of risk-communication scholarship is that members of the public tend to form opposing beliefs about technological and environmental risks on the basis of diverse cultural dispositions. Thus, on matters as diverse as climate change or silicone breast implants, ordinary people form strong, and instantaneous, emotional reactions that thereafter color how they interpret hard empirical evidence. People with *egalitarian* values will tend to blame industry for social inequities and thus will find it congenial to believe that such activities also endanger the environment and public health; in contrast, people with *individualistic* values will tend to be skeptical of asserted environmental and technological risks because accepting such assertions would justify governmental restrictions on industry, an activity that such persons admire.

Over the past two years three CCP/PEN societal studies have examined the cultural bias toward nanotechnology risks in the US public—a survey experiment of a diverse sample of some 1800 Americans—to understand whether and *how* cultural bias may affect public opinion towards nanotechnology [23]. The studies concluded that the two most promising strategies for promoting public receptivity to sound scientific information and avoiding a fractious climate are:

1) message framing
2) contextualization.

8.5.1
Cultural Message Framing

One study found that how individuals interpret information on nanotechnology risks varies depending on whether the emphasized application of nanotechnology affirms or threatens their cultural values. If the use of nanotechnology in consumer goods is made salient, for example, persons with pro-commerce individualistic values see greater benefits, but persons with anti-commerce egalitarian and communitarian values see greater risks. In contrast, if the use of nanotechnology to monitor pollution emissions is emphasized, the opposite pattern occurs. This finding suggests that *were it possible* to design message framings that simultaneously affirmed all groups' values and threatened no one's, individuals of diverse cultural outlooks would uniformly attend to information in an open-minded way.

8.5.2
Contextualization

Public perception of nanotechnology is influenced by cultural cognition. To bridge this gap between theory and practice, future research on nanotechnology risk communication should be integrated into research projects of nanotechnology scientists whose interest in effective communication of their own research can supply the setting for field experimentation. In this way, scientists involved in the study of nanotechnology and scientists involved in the study of nanotechnology risk communication can make reciprocal use of their proximity to one another to advance their common ends.

In conclusion, scientific experts are certain to play a key role in shaping perceptions of nanotechnology risks. But the second CCP/PEN study found that how members of the public are likely to react to what such experts tell them will turn *less* on what the experts actually know than on what *values* the experts are perceived to have. Yet when those same arguments were attributed to fictional policy experts, another group of subjects tended to adopt the views of the expert whose perceived values were similar to their own, and to reject the views of experts whose perceived values were different from their own, no matter what position those experts took on nanotechnology. Cultural cognition, in other words, can also generate polarization by influencing how credible people find risk communicators to be. Nevertheless, the same dynamic can be used to mitigate cultural polarization: when exposed to experts whose perceived values were like their own on both sides of the debate, subjects of diverse values tended to converge in their views.

References

1 Lubick, N. and Betts, K. (2008) *Environ. Sci. Technol.*, **42**, 3910.
2 Elder, E., *et al.* (2006) Translocation of inhaled ultrafine manganese oxide particles to the central nervous system. *Environ. Health Perspectives*, **114 (8)**, 1172–1179.
3 Poland, C.A., *et al.* (2008) *Nat. Nanotechnol.*, **3**, 423.
4 Wiesner, M.R. and Bottero, J.-Y. (eds) (2007) *Environmental Nanotechnology*, McGraw Hill.
5 Albrecht, J.M.A., Evans, C.W., and Raston, C.L. (2006) Green chemistry and the health implications of nanoparticles. *Green Chem.*, **8**, 417–432.
6 Banaszak Holl, M.M. (2009) Nanotoxicology: a personal perspective. *Wiley Interdiscip. Rev. Nanomed. Nanobiotechnol.*, **1**, 353.
7 Murphy, C.J. (2008) Sustainability as an emerging design criterion in nanoparticle synthesis and applications. *J. Mater. Chem.*, **18**, 2173.
8 Gill, V. (2009) Nano-regulation creeps closer. *Chem. World*, **6 (4)**, 10.
9 Kaegi, R., *et al.* (2008) Synthetic TiO2 nanoparticle emission from exterior facades into the aquatic environment. *Environ. Pollut.*, **156**, 233.
10 Meng, H., Xia, T., George, S., and Nel, A.E.A. (2009) Predictive toxicological paradigm for the safety assessment of nanomaterials. *ACS Nano*, **3**, 1620–1627.
11 Hutchison, J. and Greener, E. (2008) Nanoscience: a proactive approach to advancing applications and reducing implications of nanotechnology. *ACS Nano*, **2**, 395–402.
12 Lubick, N. (2009) *Environ. Sci. Technol.*, **43**, 1247.
13 Colvin, V.L. (2003) *Nat. Biotechnol.*, **21**, 1166.
14 Sweeney, S.F., Woehrle, G.H., and Hutchison, J.E. (2006) *J. Am. Chem. Soc.*, **128**, 3190.
15 Brayner, R., Vaulay, M.-J., Fievet, F., and Coradin, T. (2007) *Chem. Mater.*, **19**, 1190.
16 Maynard, A.J. (2007) Building a Safe Nanotechnology Future, Project Syndicate, http://www.project-syndicate.org/commentary/maynard1 (accessed September 30, 2009).
17 Thomas, T., Bahadori, T., Savage, N., and Thomas, K. (2009) Moving toward exposure and risk evaluation of nanomaterials: challenges and future directions. *Wiley Interdiscip. Rev. Nanomed. Nanobiotechnol.*, **1**, 426.
18 Woodrow Wilson Center (2009) Project on Emerging Nanotechnologies. Consumer products inventory, http://www.nanotechproject.org/inventories/consumer/ (accessed July 2009).
19 Wiesner, M.R. (2005) Towards a Green Nanotechnology, http://www.project-syndicate.org/commentary/wiesner1 (accessed September 30, 2009).

20 Monteiro-Riviere, N. (2006) 2006 progress report: evaluating nanoparticle interactions with skin, http://cfpub.epa.gov/ncer_abstracts/index.cfm/fuseaction/display.abstractDetail/abstract/7178/report/2006 (accessed September 30, 2009).
21 Wiesner, M. (2004) cited in The Environmental Impact of Nanotechnology, an online discussion with nanotechnology experts Dr Mark Wiesner and Dr Nancy Monteiro-Riviere. http://www.eurekalert.org/nanotalk/20041207 (accessed October, 2009).
22 Hobson, D.W. (2009) Commercialization of nanotechnology. *Wiley Interdiscip. Rev. Nanomed. Nanobiotechnol.*, **1**, 189.
23 Kahan, D. and Rejeski, D. (2009) Toward a Comprehensive Strategy for Nanotechnology Risk Communication. Available at http://www.culturalcognition.net/storage/nano_090225_research_brief_kahan_nl1.pdf (accessed October, 2009).

9
Managing (Nano)innovation

9.1
Scholars, and not Researchers

Under the concomitant threats of hypercompetition [1] and of the sustainability crisis [2] one would argue that companies everywhere would demand ever more scientists to enhance productivity, improve the quality of their products, and, at the same time, drastically reduce emissions in the environment. Paradoxically, however, in all developed countries increasingly fewer university students choose to study science, establishing an epistemological and cultural paradox: because never in the past has science had access to such a cornucopia of communication tools (media, books, museums, internet, etc.) and never before have we assisted in such a mass diffusion of social practices that deny the value and the usefulness of science [3].

Turning to management education, we realize that business schools simply do not teach science to managers. Accordingly, for example, if one asks simple questions to management people such as, for example, "How energy is produced? How much energy do you consume? What is the cost of electricity? How can you save energy?" we will most likely get poor answers [4]. The results are well rendered by this insight of web usability expert Jakob Nielsen (Figure 9.1) [5]:

> "Nokia, Ericsson, and Motorola have many great designers and usability experts who know much more than Apple about how people around the world use mobile devices. But they don't get the backing from executives to force the network operators to prioritize user experience. Steve Job's real contribution is his willingness to bang heads together to force them to upgrade their network for the trivial reason that it affords a smooth user experience on the device."

Admittedly, if a manager were to "bang together" designers of hi-tech products or lead a process to radically enhance a company's energetic

Nano-Age: How Nanotechnology Changes our Future. Mario Pagliaro
© 2010 WILEY-VCH Verlag GmbH & Co. KGaA, Weinheim
ISBN: 978-3-527-32676-1

Figure 9.1
"Nokia, Ericsson, and Motorola" – says usability guru Jakob Nielsen – "have many great designers and usability experts who know much more than Apple about how people around the world use mobile devices. But they don't get the backing from executives." Picture reproduced from useit.com, with permission.

efficiency, he/she would need to master the scientific concepts at the basis of ergonomy and energy.

Nanotechnology, of course, is no exception. How many managers, in the business community, can critically assess the potential and the limits of nanotechnology?

The simple fact that most hi-tech products are ill-designed and difficult to use, and that energy is produced and consumed according to obsolete models, are only two outcomes of the *methodological choice* of excluding science from the educational program of managerial élites worldwide.

By the same token, scientists and designers need better skills on how to communicate effectively the relevance of their work, both inwardly to management and outwardly to the public.

In brief, the present situation calls for a profound change both in the organization of intellectual work and in the scientific profession, getting back to a fertile past when there were only "scholars", and not "researchers," whose activity consisted not only in *doing research* but also in *teaching, disseminating* and *applying science* [6].

Our thesis is that renewing the education of both scientists and managers by closing the "two-cultures" gap we will enter a new era of creative work in which science and management will eventually be allied, under the unifying umbrella of *culture*. Then, thanks to their expanded educational background, managers and scientists will be able to work together to face the enormous environmental, economic, and social challenges that face our companies and, more generally, our societies.

9.2
Renewing Management and Scientific Education

The education of international management takes place in public and private prestigious schools: Yale and Harvard in the USA, INSEAD and ENA in France, the IMD in Switzerland, the London School of Economics and the "Säid" School at Oxford in the UK, to mention just a few. Invariably, the programs offered to the clientele have in common the absence of science: no physics, chemistry, or biology, and very little mathematics and engineering, are included in the curricula of contemporary managers. As educators we should thus ask ourselves: Why is science not taught to managers in our universities? And why is management not taught to scientists?

Accordingly, managers who graduate from these schools cannot critically assess—and thus effectively dominate—science and the technology issues. What eventually happens is that innovation, energy, and environment matters get delegated to chief technology officers of engineering background, with results that most often are those mentioned above by Nielsen, or those noted by Carlos Ghosn, the celebrated top manager of Nissan (and today of Renault), upon his successful restructuring of the Japanese carmaker:

> "Nissan nearly foundered because its designers were forced to take orders from engineers who knew only performance and managers who knew nothing about their customers. As a result, most of the cars the company produced may have been hot under the hood, but they were tepid in the showroom" [7].

From an even more fundamental viewpoint, we must ask ourselves: How can managers recognize that the natural sciences take part in *making sense* of ourselves and our actions, and thus establish strategies based on this awareness, if they do not *know* natural sciences?

Now, epistemologist Gloria Origgi calls for a "third culture" to expand the curriculum of our ruling classes [8]. Yet, the very idea of the existence of "two" or "three" cultures is self-contradictory, as there is only one culture made of links through its parts, which naturally include *ethics*. Khurana has shown in a recent major study [9] (Figure 9.2) that business schools today act merely as purveyors of a product, the MBA, with students treated as consumers. The professional and moral ideals that once animated and inspired business schools weakened and, he concludes, we should not thus be surprised at the rise of corporate malfeasance.

We agree with Khurana that the time has come to rejuvenate intellectually and morally the training of our future organizational and business leaders. The original ideals can and should be restored by expanding them to include science and ethics in management education. Coupled to a solid

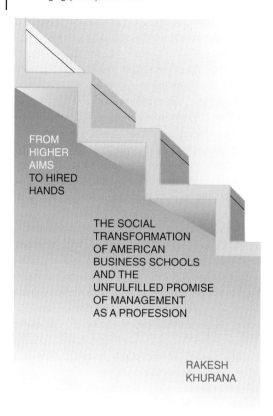

Figure 9.2
Harvard Business School professor Rakesh Khurana has shown how university-based business schools retreated from their original ideals, leaving a gaping moral hole at the center of business education and thus in management practice. Image reproduced from princeton.edu, with permission.

education in human and social sciences, a broader scientific education would allow organizational leaders to manage the change process not as a mere technical fact but as an eminently social and human process. In this manner, the executives of tomorrow will be able to integrate in the production of goods and services those human and social factors that continue to be largely neglected.

Similarly, a new broader cultural education will offer scientists the resources necessary to face the risks of ultra-specialization; and, above all, it will enable scientific professionals to fight the social emargination that everywhere is putting at risk not only the financial support of scientific research but rather the very *sense* of the scientific enterprise [3].

A closer look at the historical development of business schools [9] shows that it has been analogous to that of scientific schools: specialization and division of the field into sub-disciplines; and exactly as labor division and fragmentation of competences in industry have led to enormous productivity gains until the current crisis, so has the education of managers in business schools come to face today's legitimacy crisis:

> "Chief executives pay little attention to what business schools do or say. As long ago as 1993, Donald Hambrick, then president of the US-based Academy of Management, described the business academics' summer conference as 'an incestuous closed loop', at which professors 'come to talk with each other'. Not much has changed. In the current edition of *The Academy of Management Journal*, Rita Gunther McGrath of Columbia business school says: 'Most of what we publish isn't even cited by other academics'" [10].

Now, the task of a scientist is to produce new knowledge. Applications traditionally concern technology. Often, furthermore, the qualities of a scientist are seldom those of a good manager. Nonetheless, starting with biochemists in the late 1970s, and now including scientists from all disciplines, researchers are increasingly turning into wealthy entrepreneurs (and managers) in a global trend that is leading to the formation of a new, globally integrated elite that consists of Indian pharmaceutical and Brazilian biotechnology entrepreneurs, the Israeli chemical industry, and China's photovoltaics tycoons.

People like mathematicians Brin and Page who established Google in the late 1990s, the chemist Swanson who founded Genentech in the 1970s, and physicists Richard Friend and Zhengrong Shi who, respectively, founded the polymer electronics firms Plastic Logic in the UK and the photovoltaics giant Suntech in China, have all based their success in the marketplace on advanced scientific education. Very often, many of these scientists pursued a degree in business administration. Bob Swanson, for example, was introduced to the spirit of high tech entrepreneurship at MIT by his mentor, Richard S. Morse, through a course called "New Enterprises," a course that is still taught at MIT today.

Extending this approach, we need to rethink scientific education to include those elements of history, philosophy, sociology, and economics that are nowadays indispensable resources of the scientific profession: "Before willing to correct the deficiencies of profanes scientists must recognize their cultural gap and add these topics to their studies for a better understanding of the public" [3].

Now, science, as brilliantly shown by Feyerabend, is a collage, and not a unified system, with plenty of components derived from distinctly "non-scientific" disciplines that are often vital parts of the progress science has made [11]. In other words, the historic development of scientific disciplines has led to an open collection of knowledge in which natural sciences are becoming increasingly useful to social and human disciplines, and vice versa. A look, for instance, at how management consultancy is actually being *done* rapidly reveals the emerging interdisciplinary approach to management [12]:

> "Yamashita employs a varied toolbox of specialists, linguists, anthropologists, and artists, first to help a company define its purpose—and then to communicate that purpose both inside and out. He generally starts an engagement at the top, holding a summit with senior execs to figure out how the system works, and where it has gotten bogged down ... Once the problems have been identified, the company uses the consultancy's background as designer and marketing experts to define a solution and then moves downstream, bringing people on board by hammering the message home with creative training sessions. The role of design becomes paramount, as the consultancy produces a document, a film, or some form of media that expresses the vision".

The relevance to management of results of "in field research by anthropologists" testifies the fact that human factors—that is, people—are the ultimate judges that ensure the success of every good in the market, where the technical excellence of an invention matters less than the economic willingness of the customer to buy and to *use* it. And this, in practice, has changed the way innovation is being carried out. Nanotechnology pioneer Ray Kurzweil explains it:

> "Step one is to write the advertising brochure. This can be a real challenge. It compels you to list the features, the benefits, and the beneficiaries. You will find this impossible to accomplish if your ideas are not well formed."

> "Step two: use this brochure to recruit the intended users. If these beneficiaries don't immediately get excited about your concept, then you are probably headed down the primrose path. Invite them to participate in creating the invention. After all, if they want it so badly, let them help you invent it." [13]

This approach clearly requires the ability to *use* good language and culture as crucial driving forces to appeal to non-science practitioners, abandoning the idea that meaningful change can come only from the academic élite. When Hans Rosling (Figure 9.3), a professor of international health in Sweden, had to present global health statistics data to policy makers to convince them to invest new resources or take new decisions affecting health, he developed a highly effective program for the dynamic presentations of statistics.

Figure 9.3
Hans Rosling discussing statistical data using his Gapminder software. Image reproduced from gapminder.org, with permission.

9.3
Nexus of the Sciences

Nanotechnology is a cross-disciplinary science in which innovation occurs at the edge, in the interstices between formal academic disciplines, aiming at the production of multifunctional materials with the potential to emulate much of the functionality associated with living systems such as the active transport and transformation of matter and information, and the transduction of energy into different forms [14]. Managing innovation in nanotechnology basically requires the simple understanding that nanotech, as put by leading nanotechnology venture capitalist Steve Juvertson (Figure 9.4), is the nexus of the sciences:

> "Quite simply, it is in the richness of human communication about science. Nanotech exposes the core areas of overlap in the fundamental sciences. Nanoscale science requires scientists to cut across scientific languages to unite the isolated islands of innovation" [15].

The reason for which, then, the advent of nanotechnology should be taken as an occasion to rethink and reform scientific education is vividly illustrated by one of the responses to the above post in the blog of Juvertson:

> "Hi Steve. It is a good post. It would've been great if you could've written more about how you think the 'business side' of nanotech would differ from that of other innovations. From my personal experience (I worked on Artificial Muscles and mono-molecular thin

Figure 9.4
A former R&D Engineer at Hewlett-Packard, where seven of his communications chip designs were fabricated, Steve Jurvetson is now one of the world's leading venture capitalist investing in nanotechnology. Image reproduced from Draper Fisher Jurvetson, with permission).

films for a couple of years), the biggest hurdle was *branching out from the academic herd* towards commercializing the innovation." [15].

The reason for this difficulty, namely that of "branching out of the academic herd," has to do with the scientific deculturization invoked by Lévy-Leblond, which impacts the relationship between the scientific world and society at large.

One century after the introduction of *The Principles of Scientific Management*, the usefulness of contemporary science findings to the practice of management is almost entirely different from the mechanical view of the company as a machine suggested by Taylor in 1909 [16]. Modern control and variation theories – as shown for instance by the work of John Seddon (Figure 9.5) – would teach managers that their traditional means of control in fact give them *less* control. Similarly, instead of the principles of engineering applied to the workplace preached by Taylor, we need to discover the richness of incompleteness, and how to manage the richness of the intrinsic complexity of human systems.

Perhaps surprisingly, many of these newly educated managers would learn from systems thinking that "management by numbers" causes suboptimization, or that their view of the organization is *conditioned* by the data they use:

> "A systems view of organisations shows the fallacy of conceptualising performance problems as people problems ('if only they would do it'). They should not be considered separately from other 'task' features. Failures in co-operation, poor morale and conflicts in our organisations are symptoms, their causes lie in the system. Training in teamwork or co-operation will only treat the symptoms. The causes usually remain" [17].

Figure 9.5
John Seddon: "Management by numbers causes sub-optimization." Image reproduced from hull.ac.uk, with permission.

Companies indeed can be seen as "far-from-equilibrium" organizations that self-organize to respond continuously to change, whereas structures and solutions are temporary [18]. Since the late 1980s the theory of nonlinear dynamics known as "complexity science," and its central principles of emergence and self-organization, has been applied to understanding organizations. Concepts and examples were drawn from this branch of physics, and ten years ago one could read of those willing to learn the use of systems thinking that:

> "Managers need to have a basic grounding in business physics... managers need to become scientists of their own organizations... and develop a language about organisations that is the basis for individual and organisational learning" [19].

However, there is not and there will not be any "business physics" for the simple reason that human beings that work in companies simply do not conform to the laws of physics. To assume the existence of a "business physics" would be a vulgar error of reductionism of no practical help.

For example, the study of control and its fundamental principle that *feedback* can illuminate many things that are relevant to any organizational leader has been known at least since the early 1950s when Arnold Tustin started to explain how feedback underlies all self-regulating systems, not only machines but also the processes of life and the tides of human affairs [20]. Yet, control theory is neither a branch of physics nor of medicine (Figure 9.6) but rather it is a discipline growing from its own intellectual roots, quite independent of engineering, physiology, or economics, about all of which it nevertheless has things to say [21].

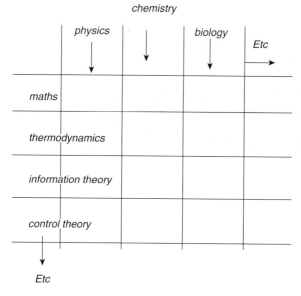

Figure 9.6
Science organizes in autonomous, reciprocally useful domains Reproduced from Reference [21], with permission.

By the same token, even if complexity recognizes change as the organizing force it does not adequately explain how *novelty* arises in organizations and, especially, the role of managers – namely of individuals – in the emergence of such novelty [22]. In other words, there is simply no need to extend concepts from one scientific field supposed to be more basic to another at another level to enhance its "scientific" credibility. Similarly, the linear model according to which science develops in a linear fashion, following a cumulative and natural progress of knowledge, has failed. The emergence of nanochemistry in the 1990s is just another example of a trend generally observed in many disciplines.

9.4
In Praise of Scientific Culture

Often promoted with sensation-seeking claims, according to which nanotechnology should have impacted

> "life science, information technology, energy-related sciences, and materials science through such applications as novel drug delivery

Figure 9.7
Eminent epistemologist Jean-Marc Lévy-Leblond (right) states that "Before willing to correct the deficiencies of profanes scientists must recognize their cultural gap and understand the public."

methods, tiny motors for robotics/medicine, new computer processing and memory platforms and highly efficient solar panels",

the first commercial products on the market were in fact stain-resistant pants, scratch-proof paint, and bouncier tennis balls [23].

Instead, it has been the extension of the old chemical methodology to nanoscale science that, eventually, is delivering the first nanotech products for advanced applications of large socioeconomic impact. This progress required the ability to look back at chemistry and at science and technology *past*. In other words, once again in agreement with Lévy-Leblond's (Figure 9.7) insight, this shows how in practice if we want to take new initiatives aimed at new objectives, we need to gain a better understanding of our history:

> "The example recalled by Paolo Galluzzi about the great Encyclopedia from the end of 18th century is fairly emblematic: right in the middle of a new scientific revolution – involving chemistry – its main supporters had to reconsider the history of their discipline. There is nothing backward-looking in such an interest in history: quite the contrary, it expresses a desire to take stock of the situation, to understand where one is in order to progress more lucidly" [6].

Researcher training models based on studying contemporary science alone are obsolete. Thus, the aim of our efforts must be to reinsert science into culture, restoring the ability to express and develop organic links between all dimensions of human activity. Science must be put back into the heart of culture and, to this end, it must come to terms with its history. In this sense, nanotechnology can and should represent a new form of Renaissance in which, in place of specialization, fragmentation, and hierarchization, science returns to be organically linked to culture.

In brief, in place of ultra-specialized researchers exclusively dedicated to the production of new knowledge, we need to educate and shape *scholars*, namely scientists who will be researchers – and thus producing new knowledge – as well as teachers and communicators who share and valorize knowledge:

> "The first step is to define new ways of training scholars. Paradoxically such new approaches could be defined on the basis of old practices from other sectors. For how can we train new professional scientists without providing them with a basic understanding of the history of science – and above all of their discipline – as well as of philosophy, sociology and the economics of science?" [3]

Now, if we take a look at the contents of any of the (best) journals in nanotechnology (e.g., *Nanotechnology*, which started publication in 1989) and ask what kind of posterity each article has had, we have to admit that most papers have not left any interesting legacy. Similarly to what happens in most other scientific disciplines, two-thirds of scientific articles are never quoted, whereas articles that are actually cited have quite short-lived visibility, as bibliographies of primary scientific literature go back on average four-five years, rarely more. In the last three years all major scientific publishers have launched journals (Figure 9.8) devoted to nanotechnology (*Nature Nanotechnology* from Nature Publishing Group, *Small* from Wiley-VCH, *Nanoscale* from RSC Publishing, *Nano Today* from Elsevier, etc.). Yet, the academic era of "publish or perish" is over, and the very intrinsic cross-disciplinary nature of nanoscale science requires the shaping of scholars whose ultimate aim is not to get the next paper published or grant approved but rather to conceive and develop breakthrough technologies and bring them to the market, involving the brightest young minds in the process.

9.5
Communicating Nanochemistry

In 2005, admitting that nanotechnology had "yet to fulfill its ultimate potential" in a perspective view of how nanotechnology will affect people in the near future, Edwards offered a set of three nanotech challenges to be achieved by 2025 [24]:

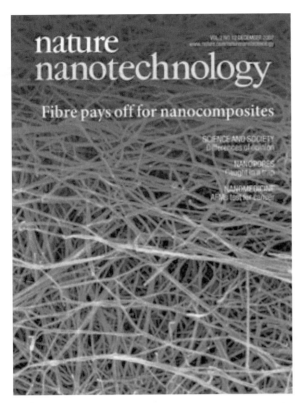

Figure 9.8
All academic publishers have launched nanotech journals. But the academic era of "publish or perish" is over. Reproduced from nature.com, with permission.

1) providing renewable clean energy,
2) improving health and longevity,
3) healing and preserving the environment.

Meanwhile, Merrill Lynch closed its "Nanotech Index" tracking the young industry's trajectory at the Stock exchange shortly after its introduction in 2004; the same bank was saved from bankruptcy in 2008 when it was taken over by a US Government-financed acquisition by another bank [25]. It is perhaps appropriate, then, to look back at past decades, remembering the promises made by scientists, to stop us from making unrealistic promises, which eventually change for the worse the attitude of the public. Again, physicist Lévy-Leblond explains the concept:

> "Society remembers promises made in the past, especially when they are not kept. In particular, it is interesting to review what physics promised in the 1950s and 1960s. We promised nuclear power would provide free energy for all. Basing their opinion on

experts' views, popular magazines of the time seriously predicted that before the end of the century everyone would have a small nuclear reactor in their own home and car (sic), and that large-scale thermonuclear fusion would be mastered. Obviously, anyone can see we are not even close to achieving these goals"[6].

One can easily realize that the declarations made by physicists in the 1950s are the same as those made today by nanotechnology advocates.

Along with Jurvetson, who has invested – and therefore put at risk – tens of millions of dollars in nanotechnology start-up companies, we believe that the fruits of nanoscale science will originate the next great technology wave. We have also provided arguments (above) that the actual development of nanotechnology will cause a paradigm shift in the scientific enterprise requiring change at all levels in its practice.

When in fact the first nanotechnology breakthroughs start reaching the market – super-strong lightweight materials, solar energy 100 times cheaper than today, methods to split water with sunlight to produce energy, improved drug delivery, fresh drinking water for all – losses will stop, profitability will jump (Figure 9.9) and along with far richer investors and companies the outcome will be a better world.

We say "world" because each of these technologies has the potential to benefit all humanity. Yet, there is a great risk in this frequent abuse of

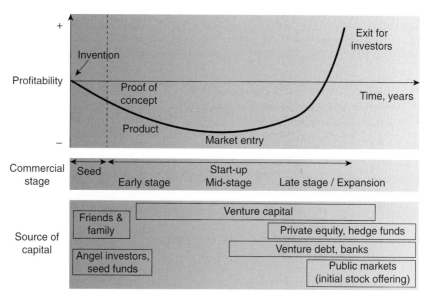

Figure 9.9
The J-curve linking return on investment in disruptive technologies to commercial stage.

Figure 9.10
David Berube, a scholar at North Carolina University, has provided precious insight into the social impact of nanotechnology. Reproduced from nanotechnologytoday.blogspot.com, with permission.

language borrowed from twentieth-century marketing as well as in referring to nano-robots, nano-machines, and other fantasy constructs, namely, *to ridicule* nanotechnology. As Berube put it (Figure 9.10):

> "What too many of us sometimes forget is that absent extensive efforts to educate the citizen-consumer, pseudo-technoliterates will people the ranks of both techno-utopians and technophobes. A failure to speak to the citizen-consumers risks fueling pervasive popular misunderstanding. Such misunderstanding could, in turn, produce formidable resistance as pseudo-technoliterates become prominent and ridicule nanotechnology."

> "The hyperbole surrounding this new technology comes not only from the media but also from scientists who exaggerate the anticipated benefits of nanotechnology to justify research funding, as well as from environmentalists and globalization opponents, who sometimes indulge in doom-and-gloom prophecies to advance their own agendas. The result is widespread misinformation and an uninformed public."

"People listen to Mander, Rifkin and even Limbaugh. In turn, their works become rallying points for technophobic dissent. On the other hand, if those who understand nanotechnology educate the citizen consumers, they may be able to mitigate many of the effects outlined above" [26].

To separate the realistic prospects from the hype surrounding this important technology, what citizens and politicians should learn in order to reach a consensus (and provide funds) is that nanotechnology is an emerging discipline with the hallmark of materials *chemistry at the nanoscale*.

A true chemistry of materials has emerged in the last 15 years as scientists from all corners of the discipline of chemistry have learned how to synthesize and exploit new types of materials from individual or groups of nanoscale building-blocks that have been intentionally designed to exhibit useful properties with purposeful function and utility. The properties of a (nano)material in fact emerge from the composition, size, shape, and surface properties of these individual building-blocks as well as self-assembled architectures made from these building blocks.

In brief, we have witnessed the inorganic, organic, polymer, and materials chemistry communities re-engineer and consolidate their skills and research interests. The boundaries that have separated these traditional chemistry disciplines in the twentieth century have broken down to create one large multidisciplinary community with a keen scientific and technological interest in "all" aspects of the chemistry of materials at the nanoscale.

The result of this evolution is that chemists are increasingly able to synthesize from the bottom-up, tailor-made (nano)materials for a myriad of applications of immense practical importance, spanning the fields of chemistry and physics, materials science and engineering, biology, and medicine.

Powerful trends are evident and in action. In research, the number of papers in the last five years has exploded, with thousands of research papers getting published yearly together with a growing complement of patents. Notably, in business, not only a growing number of start-up companies now commercialize products obtained via a nanochemistry approach but also national laboratories, military establishments, and the very big chemical companies have entered the field and joined the race for new nanomaterials. Trends in public and private funding of nanochemistry-enabled nanotechnology evolve accordingly. Global research funding in the USA, Russia (Figure 9.11), Canada, Japan, China, India, Korea, and the EU is in the range of billions of USD/Euros.

To consolidate these efforts and achieve in the next 20 years the breakthroughs that will reverse the sustainability (social, economic, and environmental) global crisis we will need newly educated managers and scientists

Figure 9.11
The Russian Government has established a €20 billion corporation to administer the funds, establish new nanotech companies, and give Russia an edge in technology in a country strong in science but weak in commercial technologies. Reproduced from rusnano.com, with permission.

whose education will be based on the integration of competences invoked in this book.

In addition, while it is true that these new managers and scientists will need to be receptive to ideas from the most disparate and apparently far domains, if we want to succeed in the change process outlined here we need to stick to simplicity and clarity, avoiding indulging in jargon since jargon is almost always the hallmark of a self-isolating community of professionals [27]. For example, we can easily realize how useful management theory, and thus its practice, has largely benefited from a discipline such as medicine by importing the concept of *prevention* and its simple but powerful message: doing things right the first time adds absolutely nothing to the cost of a product or service.

The defect that is never created cannot be missed. Identifying and eliminating the causes of problems reduces rework, warranty costs, and inspection. Hence, creating quality goods and services does not cost money, it *saves* money. Zero defects is the only acceptable performance standard; and prevention is the actual way in which quality is achieved.

Before Phil Crosby's 1979 book [28], it was commonly assumed that quality could only be achieved through inspection. Inspectors were necessary to sort the good from the bad, with ever more defect-free shipments requiring ever more examiners. With this mindset, creating quality goods and services requires increased expenditures. Crosby broke that paradigm by showing that the road to quality goods and services was through prevention, not inspection. Similarly, he suggested that management was the root cause of quality problems, and shifted the responsibility for the quality of goods and services from the quality control department to the corporate boardroom. Admittedly, all this is comprehensible and useful, whereas a sentence from a prize-winning book such as "Each value stream within the operating system must be optimized individually from end to end"[1] is neither.

---

1) Cited from a book entitled *Journey to Lean*, winner of the 2004 best management book award of Britain's Management Consultancies Association.

References

1. D'Aveni, R.A. (1994) *Hypercompetition*, Free Press, New York.
2. Elkington, J. (2007) Raising Our Game: Can We Sustain Globalization? SustainAbility Report: London. See, http://www.sustainability.com/researchandadvocacy/reports_article.asp?id=964 (accessed September 30, 2009).
3. Lévy-Leblond, J.-M. (2004) *La Science en mal de Culture*, Futuribles, Paris.
4. Armaroli, N. and Balzani, V. (2006) The future of energy supply: challenges and opportunities. *Angew. Chem. Int. Ed.*, **45**, 2.
5. Nielsen, J. (2007) Alertbox: 10 Best Intranets of 2007. Available at http://www.useit.com/alertbox/intranet_design.html (accessed October 1, 2009).
6. Lévy-Leblond, J.-M. (1997) Two cultures or none? Lecture given at the Euroscientia Conference, Rome, 1997. http://www.pantaneto.co.uk/issue8/levyleblond.htm (accessed October 1, 2009).
7. Lairmer, T. (2001) Rebirth of the Z. Time (Jan 7).
8. Origgi, G. (2006) Who is Afraid of the Third Culture? Edge. Available at http://www.edge.org/3rd_culture/origgi06/origgi06_index.html (accessed October 1, 2009).
9. Khurana, R. (2007) *From Higher Aims to Hired Hands: The Social Transformation of American Business Schools and the Unfulfilled Promise of Management as a Profession*, Princeton University Press.
10. Skapinker, M. (2008) Why business ignores the business schools. Financial Times (Jan 7).
11. Feyerabend, P. (1975) *Against Method*, New Left Books, London.
12. Reingold, J. (2004) The Liberator, Fast Company (no. 83, June), p. 82–85.
13. Kurzweil, R. (2004) Kurzweil's Rules of innovation, Technology Review, http://www.technologyreview.com/articles/print_version/kurzweil0504.asp (accessed May 2004).
14. Montemagno, C. (2003) Nanotechnology, biotechnology and complexity theory; essential tools for the engineering hybrid biotic-abiotic systems. Bio-, Micro-, and Nanosystems, 2003, ASM Conferences, 7–10 July 2003, p. 16.
15. Juvertson, S. (2004) Nanotech is the Nexus of the Sciences (Nov. 21). http://jurvetson.blogspot.com/2004/11/nanotech-is-nexus-of-sciences.html (accessed October 1, 2009).
16. Taylor, F.W. (1909) *The Principles of Scientific Management*, F.W. Taylor.
17. Seddon, J. (2003) *Freedom from Command & Control*, Vanguard Consulting, Buckingham, UK.
18. Wheatley, M. and Kellner, R. (1996) Self-organisation: the irresistible future of organising. *Strategy Leadersh.*, **24** (4), 18–24.
19. Millett, B. (1998) Understanding organisations. The dominance of systems theory. *Int. J. Organ. Behav.*, **1** (1), 1–12.
20. Tustin, A. (1952) Feedback, Scientific American, September issue.
21. Gosling, W. (1994) *Of Helmsmen and Heroes: Control Theory as a Key to Past and Future*, Wiedenfeld and Nicholson, London.
22. Stacey, R.D., Griffin, D., and Shaw, P. (2000) *Complexity and Management: Fad or Radical Challenge to Systems Thinking?*, Routledge, London.
23. Berube, D.M. (2009) The public acceptance of nanomedicine: a personal perspective. *Wiley Interdiscip. Rev. Nanomed. Nanobiotechnol.*, **1**, 2.
24. Edwards, S. (2006) *The Nanotech Pioneers*, Wiley-VCH Verlag GmbH.
25. de la Merced M.J. and Story, L. (2009) Nearly 700 at Merrill in Million-Dollar Club. The New York Times (Feb. 27).
26. Berube, D.M. (2004) The rhetoric of nanotechnology, in *Discovering the Nanoscale* (eds D. Baird, A. Nordmann, and J. Schummer), IOS Press, Amsterdam, pp. 173–192.
27. Hudson, K. (1979) *The Jargon of the Professions*, Macmillan, London.
28. Crosby, P.B. (1979) *Quality is Free*, McGraw-Hill, New York.

Index

a

abstraction 34
accumulator 49
AcP@Ag/AcP@Au 91
activation, C-H 67
active corrosion inhibitor 108
aggregation 89
alcogels 71
alcohols
– benzyl 73
– dehydration 93
– dehydrogenase 57
– oxidation 74
aldehydes 74
algae 109
alkaloids 97
alkoxysilanes 114
alloys, metal-organic 87
aluminum gallium arsenide 15
amines 33–34
amino acid chains, engineered 123
AMOLED display 114, 116
analysis
– chemical 32
– chip-based 119
Anastas, Paul 66
anti-acne drugs 127
anti-aging products 125
anti-wrinkle creams 126
antimicrobial properties 109
antireflective layer 115
appetite control 120

applications
– dental 129–132
– industrial 9, 65–85
– medical 119–137
arsenide phosphide, gallium 15
artificial muscles 159–160
asymmetric gold/silver 96
automotive catalysts 81–82
aversion, public 147
Avnir, David 41, 69–70, 87

b

bacteriostatic nanoparticles 139
Ballast SAFE 107–108
Barix encapsulation 115
batteries 10
– Li-ion 49–50
– nanotechnology-based 49–54
– NiMH 49, 52
– nLTO 52–54
beauty industry 125
benzaldehyde 73
benzene 34
benzoyl peroxide (BPO) 127
benzyl alcohol 73
Berube, David 167
bifacial configuration 16
binder, nano- 102, 104–105
bioanode 56–57
biocompatible coatings 145
biofuel cells 54–58
– stabilized enzyme 55–57

Nano-Age: How Nanotechnology Changes our Future. Mario Pagliaro
© 2010 WILEY-VCH Verlag GmbH & Co. KGaA, Weinheim
ISBN: 978-3-527-32676-1

biofuels 10, 48
– ethanol 56
– glycerol 55–57
biogels 76–81, 123–125
– self-assembly 122–123
biomarkers 139
biomineral formation 42
bioremediation 19
Biosil process 77
biotechnology 123–125
Blackberry 46
bleeding 122
boats
– hydrogen-powered electric 21
– surface protection 109
bone replacement 120
borohydride, sodium 145–146
bottom-up methodology 37
BPO (benzoyl peroxide) 127
brain, gene delivery 124–125
breast cancer, human 81
"building blocks", chemical 31–32
bulk phase, 27
buses, hydrogen fuel
 cell-powered 62
"business physics" 161

c
C-H activation 67
cage, organically modified 72
calcination 42
callogenesis 79
Canada, nanomaterial
 regulations 139–140
cancer, human breast 81
capacitors 10
capping 102–103
carbon dioxide 1–2
carbon nanotube arrays,
 multi-walled 12
catalysts
– automotive 81–82
– chiral 97
– entrapped 72

– gold nanoparticles 28
– green pharma industry 65–85
– heterogenized 69
– nano-, see nanocatalysts
– platinum-based 61–63
– recovery 66
– sol-gel 69–75
– solid 66
– two-for-one- 93–95
catalytic converter 81–82
cell compatibility 130
cells
– fuel, see fuel cells
– solar, see solar cells
ceramic honeycomb reactor 20
ceramic surfaces 133
challenges, nanotechnology
 164–165
characteristic length 11
– phenomenon-dependent 27
charge–discharge cycles 51
chemical "building blocks" 31–32
chemical formulas 35
chemical methodology 31–39
chemical protection agents
 101–117
chemical sol-gel process 106
chemical substitute 120
chip-based analysis 119
chiral metals 95–98
cinchonidine 97
cis-stilbene 94
classification, nanomaterials 140
clean energy
– renewable 164–165
– storage and supply 45–63
clean nanotechnology 139–152
Clean Power Station 23
cleaning up, water 144–145
climate change 1–2, 45
clusters 27
– nano- 132
coater, CIGS solar cells 9, 37
coatings
– biocompatible 145

lightweight materials, super-strong 166
lipase 76–77
lithium-ion (Li-ion) batteries 49–50
lithium titanate oxide (nLTO) batteries, nano-size 52–54
longevity 164–165
loop nanoantennas 13

m

macromolecular structure, hyperbranched 144–145
Magic Nano 133
Magritte 36
"management by numbers" 160–161
managing (nano)innovation 153–170
manufacturing, roll-to-roll 15–17
market, photovoltaic 5
mass market fuel cells 58
mass production, solar cells 6–9
materials
– nano-, see nanomaterials
– super-strong lightweight 166
medical tools 121
medicine
– emergency 122–123
– nano- 119–137
Mediterranean Solar Plan 4
membrane, proton exchange 61–63
mesoporous silica 42
message, cultural 150
MetaChip 78–81
metal-entrapped molecules, reactivity 90–93
metal nanoparticles, green 145–146
metal-organic alloys (MORALs) 87–91
metallic composites, enzymatically active 91

metals
– chiral 95–98
– organically doped 87–98
– surfaces 90
methane 58
methanol 95
methodological choice 154
methylated ORMOSILs 72
micelles, surfactant 42
microcapsules, sol-gel 127
microreactor, continuous 67–68
mobile and toxic nanomaterials 147
molecules, metal-entrapped 90–93
Mona Lisa 113
MORALs (metal-organic alloys) 87–91
multichannel ceramic honeycomb reactor 20
multicrystalline silicon foils 7
multifunctional nanocoatings 101–109
multifunctional textiles 109–110
multistep synthesis, one-pot 69
multiwalled carbon nanotube (MWCNT) arrays 12
Murphy, Cathrine 141
muscles, artificial 159–160

n

nano-engineering 29
nano-size lithium titanate oxide (nLTO) batteries 52–54
nano-TiO_2 142–143
nanoantennas 11–12
nanobinder Col.9 102, 104–105
nanocatalysts
– abating polluting emissions 81
– palladium 83
– porous 67
nanochemistry 27–43
– communication 164–169
– medical applications 119–137

green chemistry 65–85
– metal nanoparticles 145–146
– nanomaterials 141
– nanotechnology 143–146
– "12 principles" 65–66

h
health 164–165
hemostasis 122–123
heterogenized catalysts 69
High Tech Nature (HTN) 78–79
high-throughput printing technologies 6
historical hysteresis 30
honeycomb reactor, multichannel ceramic 20
hormone therapy 120
Hot-Module 59
human breast cancer 81
hybrid cars 48
hybrid coating 108
hydrocarbons, oxidation 67
hydrogen
– electrochemical oxidation 59
– electrolysis 19, 22
– solar 19–24, 47
– storage 10
hydrogenation 82
hydrogen-powered electric boat 21
HydroLAb 19, 60
hydrophobicity 103
Hydrosol project 21
hydroxide ions 92
hyperbranched macromolecular structure 144–145
hypercompetition 153
hysteresis, historical 30

i
immunosuppressant 120
indium gallium phosphide 15
individualistic values 149

industrial applications
– green pharma industry 65–85
– nanoscale 9
inflammation 130
ink, nanostructure CIGS 9, 37
inner porosity 70
innovation management 153–170
inorganic UV-protector 112
insulation, thermal 10
Intergovernmental Panel on Climate Change (IPCC) 1
internal combustion engine 45–49
iPhone 46
isophorone 98

j
J-curve 166
Juvertson, Steve 159–160

k
Kekulé 33
keto-olefins, cyclic 98
ketones 33–34
Khurana, Rakesh 155–156
kinetics, fast 72
Kuhn, Thomas 41
Kurzweil, Ray 158

l
Lavoisier 32–33
LCDs, flexible 116
length scales, characteristic 11
Lévy-Leblond, Jean-Marc 163–164
life cycle, nanomaterials 146–147
$LiFePO_4$ nanoparticles 51
light
– protection against 111–117
– ultraviolet 111, 128
lighting, solid state 10, 101, 116–117

electrolysis hydrogen 19, 22
elimination 33–34
emergency medicine 122–123
emergent properties 27
emerging nanotechnologies 149
emissions, polluting 81
emulsion–solution–transfer (EST) process 75
enantiomers 98
encapsulation, Barix 115
energy
– clean 45–63, 164–165
– conservation 101–117
– conversion, storage, and conservation technologies 10
– renewable clean 164–165
– solar 1–25
energy carriers, characteristic length and time scales 11
engineered amino acid chains 123
entrapment 89
– catalysts 72
environment protection 164–165
– US EPA 40
environmentalists, "skeptical" 5–6
enzymatically active metallic composites 91
enzymes
– cytochrome P450 80–81
– lipase 76–77
– oxidoreductase 56
– stabilized 55–57
epoxy-silica hybrid coating 108
epoxyfunctionalized alkoxysilanes 114
ethanol, biofuel 56
EV1 47
exposure potential, nanomaterials 146

f
fast kinetics 72
FDA regulated products 120
"feed-in" tariff schemes 5–6

Feynman, Richard P. 37–40
films
– crosslinked silicone 102–103
– polymer 15
– thin, see thin films
– thin-film photovoltaics 6
fission, nuclear 1
flexible LCDs 116
flexible solar modules 18
foils, multi-crystalline silicon 7
formaldehyde 95
fragrances 73–74
framing, cultural message 150
fuel cells 10
– biological 54–58
– buses 62
– Hot-Module 59
– mass market 58
– PEM 61–63
– portable 20
functional randomness 41
functionalization
– epoxy- 114
– surfaces 106
funding, nanomedicine 136

g
gallium arsenide phosphide 15
Gapminder software 159
gels, bio- 76–81, 123–125
gene delivery 124–125
generalization 34
Ghosn, Carlos 156
glass
– biogels 76–81
– self-cleaning 142–143
– surfaces 133
global warming 1–2, 45
glycerol 55–57
gold, asymmetric 96
gold clusters 27–28
goods, protection 110–111
graphite 53–54

- epoxy-silica hybrid   108
- nano-   101–109
- ORMOSIL-based materials   107
- sol-gel   106
cognition, cultural   148–151
Col.9 (nanobinder)   102, 104–105
color retention   105
combustion engine,
   internal   45–49
commercialization, nanomedicine
   136
Common Lilac (*Syringa vulgaris*)
   80
communication
- nanochemistry   164–169
- nanotech risks   148–151
complexity science   161
concentrated solar power (CSP)   3
conductive polymer matrix   55
Congo Red   94–95
conservation, energy   10, 101–117
"container construction" design
   22
contextualization   150–151
continuous microreactor   67–68
conversion technologies, energy
   10
converter, catalytic   81–82
Cool Pearls™   127
copper-indium-gallium-selenide
   (CIGS) solar cells   8–9, 37
corrosion inhibitor   108
cosmetics   125–129
creams, anti-wrinkle   126
credibility, scientific   162
Crosby, Phil   169
crosslinked silicone film   102–103
Cultural Cognition Project (CCP)
   148–151
cultural heritage, protection
   110–111
cultural message framing   150
culture
- scientific   162–164
- "two-cultures" gap   154

cyclic keto-olefins   98
cytochrome P450   80–81
cytotoxicity   130

d

dehydration, alcohol   93
dehydrogenase, alcohol   57
dendrimers   144–145
dental applications   129–132
design criteria   141
diabetes   66
diagnostic tests   119
dipole antennas   12
Dirac, Paul   32
discharge, charge–discharge
   cycles   51
DNA delivery   124–125
doped materials
- organically   87–98
- sol-gel   69
- water   121
drugs
- anti-acne   127
- delivery systems   119
dye
- dye-sensitized solar cells   16–17
- organic   88
- Rhodamine   112–113

e

education, scientific   155–162
efficiency, light-to-electricity
   conversion   13–14
electric boats, hydrogen-powered
   21
electric energy storage   2
electric SUV   52–53
electric vehicles   47
electrical energy   1, 45
electrocatalysts, platinum-based
   61–63
electrochemical oxidation,
   hydrogen   59

nanoclusters 132
nanocoatings, multifunctional 101–109
nanocrystallites 94
nanofiller composite 129–132
nanoinnovation management 153–170
nanomaterials 30, 39–43
– classification 140
– exposure potential 146
– life cycle 146–147
– regulations 141–143
– self-assembly 41
nanomedicine 119–122
– commercialization 136
– funding 136
nanoparticles
– bacteriostatic 139
– clearance 130
– green metal 145–146
– LiFePO$_4$ 51
– organically modified silica 125
nanophysics 29
nanoporous catalysts 67
nanoscience, solar power/energy 9–15
NanoSelect platform technology 82–84
nanosols 106
"nanosome" particles 125–126
nanostructures
– CIGS ink 9, 37
– platinum 63
nanotechnology 30
– challenges 164–165
– chemical approach 29–31
– cosmetics 125–129
– emerging 149
– green 143–146
– orthopedics 129–131
– risks 148–151
– social impact 167
– sustainable 139–152

nanotechnology-based batteries 49–54
nanotubes, carbon 12
nanowires, single-crystal 62
natrium, see sodium
Neowater 120–121
"New Enterprises" 157
nickel metal-hydride (NiMH) batteries 49, 52
Nielsen, Jakob 153–155
noise-less submarine engines 45
non-scalable regime 27
non-viral vector 125
non-wetting templates, particle replication 134–135
nuclear fission 1

o

octamers, gold 27–28
oil-free alternatives 47
OLED lighting, solid-state 101, 116–117
olefins, keto- 98
one-pot multistep synthesis 69
orange windows, semitransparent 17
organic dye 88
organic shell, protective 84
organically doped metals 87–98
organically modified cage 72
organosilica 125
– hybrids 104
– thin films 113
ORMOSILs 71–73, 75–77
– coating materials 107
– medical applications 124–125
orthopedics 129–131
oxidation
– alcohols 74
– electrochemical 59
– hydrocarbons 67
– traditional processes 73
oxidoreductase enzymes 56
Ozin, Geoffrey 29

p

P450 enzymes, cytochrome   80–81
paint, scratch-proof   163
palladium nanocatalysts   83
PANI (polyaniline)   93
pants, stain-resistant   163
paradigm shift   41, 166
particle replication in non-wetting templates (PRINT)   134–135
perruthenate, tetra-*n*-propylammonium   71–73
pharma industry, green   65–85
phenylfunctionalized alkoxysilanes   114
"philosopher's stones", sol-gel catalysts   69–75
phosphide, gallium arsenide/indium gallium   15
photocatalysis   10
photovoltaics
– market   5
– thin-film   3–4, 6
physical protection agents   101–117
physics
– "business"   161
– nano-   29
pinacol rearrangement   93
platinum-based electrocatalysts   61–63
platinum nanostructures   63
plug-in hybrid cars   48
pollution   45, 81
polyaniline (PANI)   93
polycondensates   114
polyester sieves   110
polymer films   15
polymer matrix, conductive   55
polymer technology   116
polysilicon   7
poly(styrene sulfonic acid)   89
polyurethane   103
poly(vinylbenzyltrimethylammonium hydroxide) (PVBA)   91
porosity, inner   70
portability, electrical energy   45
portable fuel cells   20
power, solar   1–25
– solar, *see also* photovoltaics, solar cells
power density   50–51
Power Plastic®   16
power windows   16
Prague Castle   110–111
precautionary principle   141
PRINT (particle replication in non-wetting templates)   134–135
printing technologies, high-throughput   6
Prius, Toyota   48
processes
– Biosil   77
– chemical sol-gel   106
– emulsion–solution–transfer   75
– green nanotechnology   143–144
– PRINT   134–135
– traditional oxidation   73
product pyramid   116
products
– anti-aging   125
– FDA regulated   120
– green nanotechnology   143–144
protection
– environment   164–165
– organic shell   84
– physical/chemical agents   101–117
proton exchange membrane (PEM) fuel cell   61–63
pseudo-technoliterates   167
public aversion   147
"publish or perish"   165
PV, *see* photovoltaics

q

quotations   164

r

randomness, functional 41
rapid toxicity testing 81
REACH (Registration, Evaluation, authorization and Restriction of Chemicals) regulations 142
reactivity, metal-entrapped molecules 90–93
reactors
– continuous microreactor 67–68
– honeycomb 20
rearrangement, pinacol 93
recovery, catalysts 66
Reetz, Manfred 76
regulations, nanomaterials 141–143
renewable clean energy 164–165
representation 35–36
research brief 149
restorative nanofiller composite 129–132
retinol 126
Revitalift 126
[Rh]@Ag 94
Rhodamine dye 112–113
risks, nanotechnology 146–151
roll-to-roll manufacturing 15–17
Rosling, Hans 158–159
RUSNANO 168

s

SAFE 107–108
Schrödinger equation 34
scientific credibility 162
scientific culture 162–164
scientific education 155–162
scientific revolution 41
scratch-proof paint 163
sealing, surfaces 133
SEBC (stabilized enzyme biofuel cell) 55–57
Seddon, John 160–161
self-assembly 41
– biogels 122–123

self-cleaning 103–104
– glass 142–143
semitransparent orange windows 17
sieves, polyester 110
SilFoil modules 7
Silia*Cat* 73
silica
– epoxy-silica hybrid coating 108
– mesoporous 42
– organically modified nanoparticles 125
– surfaces 68
silicate 42
silicon foils, multi-crystalline 7
silicone film, crosslinked 102–103
siloxane backbone structure 108
silver 88
– asymmetric 96
– nanocrystallites 94
single-crystal nanowires 62
single-nanocatalyst concept 81–83
"skeptical environmentalist" 5–6
smartphone 46
social impact of nanotechnology 167
social transformation 156
sodium borohydride 145–146
soil repellency 109
sol-gel catalysts 69–75
sol-gel coating 106
sol-gel "doped" materials 69
sol-gel microcapsules 127
sol-gel process, chemical 106
solar cells
– CIGS 8–9, 37
– dye-sensitized 16–17
– flexible 18
– mass production 6–9
– translucent 16–17
– *see also* photovoltaics
solar hydrogen, from water 19–24, 47

solar power 1–25
– concentrated 3
– nanoscience 9–15
solar thermal plant 3
solar transmittance 115
solid catalysts 66
solid state lighting 10
– OLED 101, 116–117
sols, nano- 106
Solucar 3
splitting, water 20–21
spray, sealing 133
stabilized enzyme biofuel cell (SEBC) 55–57
stain-resistant pants 163
Starburst® dendrimers 144–145
cis-stilbene 94
storage
– clean energy 45–63
– electric energy 2
– technologies 10
– *see also* batteries
strain, zero strain material 53
styrene sulfonic acid 89
submarines, hydrogen fuel cell-powered 60
– noise-less engines 45
sun's energy, *see* solar power
super-strong lightweight materials 166
supercapacitor 50
supply, clean energy 45–63
surfaces
– ceramic/glass 133
– functionalization 106
– metallic 90
– protection 109
– sealing 133
– silica 68
– Zyvere 103–104
surfactant micelles 42
surgery 122–123
sustainability crisis 48, 153
sustainable nanotechnology 139–152

SUVs, electric 52–53
synthesis 32
– one-pot multistep 69
(*Syringa vulgaris*) 80

t

tariff schemes, "feed-in" 5–6
"technoliterates", pseudo- 167
templates, non-wetting 134–135
tetra-*n*-propylammonium perruthenate (TPAP) 71–73
2,2,6,6-tetramethylpiperidine-1-oxyl (TEMPO) 73–74
textiles, multifunctional 109–110
therapy, hormone 120
thermal insulation 10
thin films
– organosilica 113
– photovoltaics 3–4, 6
Thomson, D'Arcy 30, 42
time scales, characteristic 11
titania, inorganic UV-protector 112
– self-cleaning glass 142–143
toluene 73
Toumey, Chris 39–40
toxicity 130
– nanomaterials 147
– testing 81
Toyota Prius 48
"TPAP (tetra-*n*-propylammonium perruthenate) 71–73
"traditional oxidation processes 73
transfer processes, emulsion–solution– 75
transformation, social 156
translucent solar cells 16
transmittance, solar 115
"turbogas" 58
"two-cultures" gap 154
two-for-one-catalyst 93–95

u

ultra-specialization 156, 164
ultraviolet (UV) light 111, 128
Union for the Mediterranean 4
United Nations Intergovernmental Panel on Climate Change (IPCC) 1
uranium 37–39
US Environmental protection agency 40
UV-Pearls™ 128–129

v

vector, non-viral 125
VOC (volatile organic compound) 104

w

washing machine 134
waste avoidance 65
water
– cleaning up 144–145
– "doped" 121
– solar hydrogen 19–24, 47
– splitting 20–21
wavefunction 34
web usability 153–154
windows, power 16

z

zeolites 43
zero strain material 53
Zyvere surface 103–104